TM 43-0139
Painting Instructions
for
Army Materiel

This manual is published to provide information and guidance to personnel charged with painting and marking equipment for which the U.S. Army has responsibility. It contains instructions for treating surfaces to remove corrosion, and procedures for preventing corrosion by applying protective coatings. Although many paint systems are covered, special emphasis is placed on the Chemical Agent Resistant Coatings (CARC), because they are particularly effective in resisting corrosion and chemical penetration, and are also decontaminated more easily than are other coatings.

Should you have suggestions or feedback on ways to improve this book please send email to Books@OcotilloPress.com

Edited 2021 Ocotillo Press
ISBN 978-1-954285-54-5

Ocotillo Press
Houston, TX 77017
Books@OcotilloPress.com

TECHNICAL MANUAL

PAINTING INSTRUCTIONS
FOR
ARMY MATERIEL

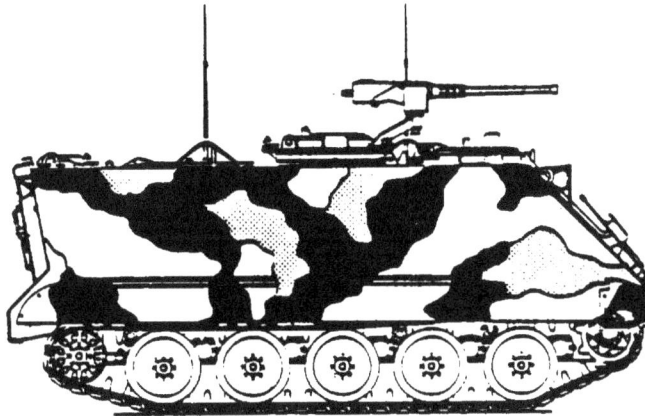

Approved for public release. Distribution is unlimited.

*This manual supersedes TM 43-0139, 1 August 1986.

HEADQUARTERS, DEPARTMENT OF THE ARMY
27 JULY 1988

CHANGE

NO. 3

**HEADQUARTERS,
DEPARTMENT OF THE ARMY
WASHINGTON, D.C.   29 FEBRUARY1996**

PAINTING INSTRUCTIONS FOR ARMY MATERIEL

DISTRIBUTION STATEMENT A: Approved for public release; distribution is unlimited

TM 43-0139, 27 July 1988, is changed as follows:

1. Remove and insert pages as indicated below. New or changed text material is indicated by a vertical bar in the margin. An illustration change is indicated by a miniature pointing hand.

Remove pages	Insert pages
i through iii / (iv blank)	i through iv
2-9 through 2-14	2-9 through 2-14
3-1 through 3-6	3-1 through 3-6
3-9 through 3-12	3-9 through 3-12
4-1 and 4-2	4-1 and 4-2
5-3 (5-4 blank)	5-3 (5-4 blank)
A-1 and A-2	A-1 and A-2
B-1 through B-9 / (B-10 blank)	B-1 through B-10
D-5 through D-8	D-5 through D-8
Index 1 and Index 2	Index 1 and Index 2

2. Retain this sheet in front of manual for reference purposes.

By Order of the Secretary d the Army:

DENNIS J. REIMER
General, United States Army
Chief of Staff

Official:

JOEL B. HUDSON
Acting Administrative Assistant to the
Secretary of the Army
01643

DISTRIBUTION:
    To be distributed in accordance with DA Form 12-34-E, block no. 0868, requirements for TM 43-0139.

CHANGE

NO. 2

HEADQUARTERS, DEPARTMENT OF THE ARMY
AND HEADQUARTERS, U.S. MARINE CORPS
WASHINGTON, D.C., 21 June 1990

**Painting Instructions
for
Army Materiel**

Approved for public release; distribution is unlimited.

TM 43-0139, 27 July 1988, is changed as follows:

1. Remove and insert pages as indicated below. New or changed text material is indicated by a vertical bar in the margin. An illustration change is indicated by a miniature pointing hand.

Remove pages	Insert pages
3-3 and 3-4	3-3 and 3-4
4-5 and 4-6	4-5 and 4-6
5-29 and 5-30	5-29 and 5-30
B-7 and B-8	B-7 and B-8

2. Retain this sheet in front of manual for reference purposes.

By Order of the Secretary of the Army:

CARL E. VUONO
General, United States Army
Chief of Staff

Official:

WILLIAM J. MEEHAN, II
Brigadier General, United States Army
The Adjutant General

DISTRIBUTION:

To be distributed in accordance with DA Form 12-34B, Maintenance Requirements for Painting Instructions for Field Use.

CHANGE

HEADQUARTERS
DEPARTMENT OF THE ARMY
WASHINGTON, D.C., 26 January 1990

NO. 1

PAINTING INSTRUCTIONS
FOR
ARMY MATERIEL

Approved for public release; distribution is unlimited

TM 43-0139, 27 July 1988, is changed as follows:

1. Remove and insert pages as indicated below. New or changed text material is indicated by a vertical bar in the margin. An illustration change is indicated by a miniature pointing hand.

Remove pages	Insert pages
iii/iv	iii/iv
2-3 and 2-4	2-3 and 2-4
3-11 and 3-12	3-11 and 3-12
B-1 and B-2	B-1 and B-2
B-9/B-10	B-9/B-10
C-11 and C-12	C-11 and C-12
Index 5 and Index 6	Index 5 and Index 6

2. Retain this sheet in front of manual for reference purposes.

By Order of the Secretary of the Army:

CARL E. VUONO
General. United States Army
Chief of Staff

Official:

WILLIAM J. MEEHAN, II
Brigadier General United States Army
The Adjutant General

DISTRIBUTION:
   To be distributed in accordance with DA Form 12-34B, Maintenance Requirements for Painting Instructions for Field Use.

Technical Manual
No: 43-0139

HEADQUARTERS
DEPARTMENT OF THE ARMY
WASHINGTON, D.C. 27 July 1988

PAINTING INSTRUCTIONS
FOR
ARMY MATERIEL

REPORTING ERRORS AND RECOMMENDING IMPROVEMENTS
You can help improve this manual. If you find any mistakes or if you know of a way to improve these procedures, please let us know. Mail your letter or DA Form 2028 (Recommended Changes to Publications and Blank Forms), or DA Form 2028-2 located in the back of this manual directly to: Commander, US Army Aviation and Troop Command, ATTN: AMSAT-I-MP, 4300 Goodfellow Blvd., St. Louis, MO 63120-1798. You may also submit your recommended changes by E-mail directly to <daf2028@st-louis-emh7.army.mil>. A reply will be furnished directly to you.

TABLE OF CONTENTS

Page

*This manual supersedes TM 43-0139, 1 August 1986

LIST OF ILLUSTRATIONS

LIST OF TABLES

CHAPTER 1

INTRODUCTION

Section I. GENERAL

NOTE
Unusual terms are defined in the Glossary located at the back of this manual.

1-1. PURPOSE

This manual is published to provide information and guidance to personnel charged with painting and marking equipment for which the U.S. Army has responsibility. It contains instructions for treating surfaces to remove corrosion, and procedures for preventing corrosion by applying protective coatings. Although many paint systems are covered, special emphasis is placed on the Chemical Agent Resistant Coatings (CARC), because they are particularly effective in resisting corrosion and chemical penetration, and are also decontaminated more easily than are other coatings.

1-2. SCOPE

a. This manual discusses materials associated with painting operations, procedures for marking and camouflaging equipment, and methods of applying paint.

b. This manual is applicable to equipment under U.S. Army jurisdiction, whether assigned to active service or in wet or dry storage. Additional information for painting watercraft and aircraft are contained in TB 43-0144, Painting of Vessels, and TM 55-1500-345-23, Painting and Marking Army Aircraft. For additional information for painting military vehicles, construction equipment and material handling equipment, refer to TB 43-0209. For detailed corrosion correction and prevention techniques, refer to TB 43-0213.

1-3. POLICY

a. There are safety, health and environmental requirements associated with all aspects of painting operations. These are outlined in Section II of this chapter, Safety Summary. Personnel must keep these requirements in mind before, during and after undertaking any painting activity. Any questions should be directed to local preventive medicine/industrial hygiene personnel.

b. Equipment with applied coatings providing satisfactory protection will not be altered solely for conformity to the requirements herein. Complete repainting should be done only when the existing finish has deteriorated to the extent that it no longer protects the underlying surface or when higher authority mandates. Camouflage patterns may be painted on items coated with Green 383 anytime after pattern design development, at commanding officer's discretion.

c. Chemical Agent Resistant Coatings (CARC) are required on all combat, combat support, and combat service support equipment. Current alkyd and lacquer paints must be removed after chemical agent exposure as the paints absorb liquid agents and release the agents over time, causing a contact hazard. Since CARC does not absorb chemical agents it does not create long term contact hazards.

d. Only Intermediate (General Support and Direct Support) and Depot level personnel with equipment and paint booths meeting OSHA standards are authorized complete painting and repainting with any topcoat or primer. Unit level personnel are permitted to use topcoats and primers for touch-up efforts only.

e. The style, size and exact location of markings prescribed in this manual are specified in applicable technical bulletins in the 43 and 746 series and other DA technical publications. Markings may be applied in the form of adhesive backed markers of the prescribed color, or may be painted on when markers are unavailable or application must be made on canvas or other porous surfaces.

f. Special markings for vehicles in administrative use are included in AR 58-1.

g. Under tactical conditions, when requirements for concealment outweigh those for recognition, all conspicuous markings may be obscured or removed by the authority and at the discretion of the major organization commander present. Protective red cross markings may be obscured only at the direction of the responsible major tactical commander.

h. Major end items and major components with exposed surfaces painted with CARC will have the word 'CARC' stenciled on them in close proximity to the data plate. Refer to para 4-9 and Chapter 6, Section II.

i. Markings on the exterior of tactical equipment will be applied using CARC in accordance with para 4-9.

j. Safety markings, including hazard warning and caution information, for non-tactical equipment, tactical not subject to the Army camouflage policy, and equipment at fixed facilities will comply with the provisions of AR 385-30. Materiel painted in camouflage requiring hazard warning and caution information will have this information applied in accordance with para 4-9.

k. Additional marking policy is contained in AR 750-1, Maintenance of Supplies and Equipment Painting, Army Materiel Maintenance Policies.

1-4. PURPOSE OF PAINTING

a. Corrosion Protection. The primary function of painting is to protect metals, wood, and other material against corrosion and decay.

(1) Paint should not be applied to unseasoned wood, since paint retards the seasoning process and fails to form a proper coating under such conditions.

(2) Certain paints adhere to a given surface better than others and therefore furnish a better protective coating. The first or base coat should penetrate into the minute depressions or pits in the material and should adhere well enough to form a good bond for any additional coats.

(3) The success of painting depends on the selection of a suitable paint, and also upon the care used in preparing the surface, which should be thoroughly cleaned, dry, and smooth. Other factors include the method of application and weather conditions.

b. Camouflage. Camouflage of Army materiel is a function of paint. Para 4-2 discusses reasons for camouflaging. Additional information on camouflage can be found in FM 5-20, TM 5-200, and FC 90-7.

c. Visibility. White and light-tinted paints are frequently used on interior surfaces to increase the visibility in spaces with limited access to outside light. In this respect, paint can serve to increase visibility with existing natural or artificial light, or it can serve to reduce the amount of natural or artificial light required in a given interior space.

d. Chemical Agent Resistance. Chemical Agent Resistant Coatings (CARC) are used to protect combat, combat support, and combat service support equipment from chemical agent penetration. These coatings can be decontaminated relatively easily.

e. Identification. Paint can be used to apply identification marks to equipment. Chapter 6 contains instructions for marking Army materiel. Markings on camouflaged equipment will be in accordance with para 4-9.

Section II. SAFETY SUMMARY

1-5. GENERAL

This section outlines safety, health and environmental requirements applicable to all painting operations. Safety and health requirements are the same, regardless of paint system used, except where specifically identified. If there is ever uncertainty as to what is required, contact local preventive medicine/industrial hygiene personnel.

a. *Vapors.* Thinners used with paints and primers may have harmful effects. Continued breathing of vapors during and after painting operations should be avoided. Toxic vapors may persist, in some cases, for many days indoors after painting operations. Every effort must be made to ensure proper ventilation of the paint area to rid the area of toxic vapors as quickly as possible. All personnel must be made aware that toxic vapors may be present. Avoid inhaling toxic vapors.

b. *Contact with Paint Materials.* Avoid skin contact with paints, primers, removers and thinners, particularly if there are cuts or open wounds on the hands. Unwashed hands may convey toxic material to food. Many paints and primers contain lead, chromium, or other toxic materials which may enter the body when paint-contaminated food is eaten. Many of the thinners are also toxic, and can enter the human body through the skin or by eating contaminated food. Personal protective equipment (PPE) as recommended in para 1-7a should be worn to prevent skin contact.

c. *Fire.* The mist that comes from a spray gun is highly flammable. A spark will cause it to flash. Smoking is prohibited in paint shops. Open cans containing paint removers, thinners, paints, and primers are a fire hazard. Empty drums or other containers in which paints, primers and thinners have been shipped are potential hazards since they often contain enough vaporized material of a flammable nature to cause explosions. Accumulated overspray in booths and in cracks and corners of the paint shop is particularly dangerous for it easily flares up. Oil or solvent-soaked cloths, if not properly contained and promptly disposed of in accordance with AR420-47 may cause fire by spontaneous combustion. Fires which occur in spray booths result from six principal causes: broken electric lamps and other electrical defects; cleaning interior of booths, fans, and motors with flammable solvents; accumulations of deposits in the booths, tubes, and vent pipes; defective fans and motors used for ventilating the booths; poorly designed vent tubes; or static electricity.

d. *Safe Air.* If it is at all practical, painting of material should be accomplished in a properly designed and operated paint booth. Adequate forced draft ventilation for indoor touchup work should be provided to carry off vapors. Respirators should be worn during all spray-painting operations. Refer to para 1-7 for specific respiratory protection required.

e. *Safe Practices.* Preparations containing benzene should not be used for spraying. Only electrical equipment/wiring conforming to NFPA Article 70 will be used where spray-painting is being done. Paints should be stored in a steel cabinet meeting OSHA requirements. Once opened, cans containing paint removers, thinners, paints, and paint materials should be covered tightly before being stored or put away overnight. Do not apply heat or flame to drums, cans, or other containers that have contained flammable materials. Obsenrve safe operating procedures at all times, particularly when handling cleaning materials. For disposal of unserviceable paints or primers, refer to para 1-9.

1-6. MATERIAL SAFETY DATA SHEETS (MSDS)

MSDS are prepared by the manufacturer and should accompany each single shipment or batch of paint, primer or thinner. It is mandatory that personnel working with these substances read this information. Because of the variations involved, MSDS must be reviewed for each shipment procured on a single purchase order. MSDS must be filed in a location readily accessible to workers exposed to the substances. MSDS also assist management by directing attention to need for specific control engineering, work practices and protective measures to ensure safe handling and use of the material. Along with the product's ingredients and specific protection information, the MSDS contain the following data:

a. *Reactivity Data.* This informs the paint user about the stability, hazardous decomposition, or polymerization properties of the coating.

b. *Spill and Disposal Procedures.* This informs the paint user the steps to be taken for proper spill or disposal methods.

c. Fire and Explosion Hazard Data. This informs the paint user about the flash point of the product, special fire fighting procedures, and the extinguishing media.

d. Health Hazards. Personnel should be familiar with emergency and first aid procedures as outlined in the product's M SDS. This includes medical procedures to be followed if the product is inhaled, or if the product has come in contact with the skin or eyes of an individual.

1-7. CONTROL MEASURES

a. Personal Protection.

(1) Personal protective equipment (PPE) used in conjunction with respiratory protection equipment (para 1-7b) during spray-painting includes cloth coveralls, eye protection, and head coverings. Cloth gloves are suitable unless cellosolve acetate (2-ethoxyethyl acetate) is present in the paint, solvent, or primer. When this solvent is present, silicon rubber gloves are recommended. Spot painters applying paint by brush or roller must wear work clothing and gloves affording full skin coverage. Persons who clean mixing and painting accessories should wear full PPE to preclude solvent absorption and defatting of the hands caused by the thinner.

(2) If a solvent with a skin notation is being used, then impervious gloves must be used. Barrier creams are useful in preventing paint from adhering to the skin and in combating the "dryness" associated with the defatting action of most solvents; however, their usefulness in preventing the absorption of solvent through skin is not documented. Solvents must never be used to remove paint/coating from the skin.

(3) Work clothing should be provided. After completing painting or sanding operations, hygienic showers should be taken prior to changing into street clothing.

b. Respiratory Protection.

WARNING
Prior to beginning any painting operation, preventive medicine/industrial hygiene personnel must be contacted. Painting materials can cause serious health problems if used improperly or without adequate respiratory protection.

(1) Before beginning painting operations, contact local preventive medicine/industrial hygiene personnel, who will determine minimum respiratory protection requirements in accordance with TB MED 514. Depending on method of application and facilities available, some sort of respirator will probably be required.

(2) Levels of exposure to contaminants will be documented by preventive medicine/industrial hygiene personnel. Additional monitoring is required whenever there has been any change in the operation which could result in new or additional exposures.

c. Ventilation. The use of respiratory protection equipment does NOT waive the requirement for engineering control measures. The ventilation design specifications for spray paint booths are in TB MED 514.

d. Preferred Coatings. Whenever available, lead-and chromate-free coatings should be used.

e. Warning Labels. Warning labels are required on products which contain materials hazardous to your health. Read these warnings.

1-8. MEDICAL SURVEILLANCE

Medical surveillance to detect adverse health effects will be determined by the installation medical authority (IMA) based on the specific constituents of the coating. In general, medical surveillance is required for anyone who works more than 30 days per year in either a paint spraying operation or in a brush or roller application when respiratory protection is required. Personnel involved in painting at direct support, general support, and depot levels of maintenance will normally require surveillance. Vehicle/equipment operators and unit maintenance section personnel usually do not perform enough brush touch-up painting to warrant medical surveillance.

1-9. DISPOSAL

a. *General.* Unusable paint mixtures, paint components, primers, thinners and other materials may be considered hazardous waste and require disposal in accordance with Federal, state, DOD, and DA hazardous waste regulations. This may apply to dried paint/primer waste as well. Consult local environmental personnel for proper disposal guidance.

b. *Method of Disposal.* The method used to dispose of this waste stream depends on the types of paint used. When the paints contain no hazardous heavy metals, the liquid portion of the waste stream may be able to be discharged into the sanitary sewer and the sludge disposed of in a sanitary landfill. The environmental coordinator should be contacted to ensure state and local ordinances are not violated. If heavy metal-based paints have been used, samples of the liquid and sludge should be analyzed for the characteristic of extraction procedure (EP) toxicity (heavy metals) and disposed of accordingly.

(1) If sanitary sewer serving a paint processing operation discharges to a government-operated sewage treatment plant (STP), notify the STP operator of the approximate additional loadings of total organic carbon, biochemical oxygen demand, and total processing operation prior to discharge.

(2) If a sanitary sewer serving a paint processing operation discharges to privately owned treatment works, pretreatment of wastewaters may be required by local regulatory authority. The installation environmental coordinator should determine such cases with public environmental regulators having primacy over government installations. Pretreatment regulations have been summarized by the USAEHA Water Quality Information Paper No. 13.

1-10. SPECIFIC HAZARDS

Listed below are the various specific hazards associated with spray painting, cleaning, fires, safety equipment, hazardous materials, temperatures and equipment.

a. *All Painting.*

WARNING

The local safety office and preventive medicine support activity must be consulted before beginning/changing any painting operation, regardless of the material used.

WARNING

Dry/cure freshly painted materials only in well-ventilated or unoccupied areas for a minimum of 30 minutes to allow solvents to flash off.

b. *Spray Painting.*

WARNING

All personnel who work in or near a spray painting booth must wear a NIOSH approved respirator as well as personal protective equipment (PPE), when spray painting operations are underway.

WARNING

Only one person will spray paint at a time unless all people are protected in accordance with para 1-7. This is to eliminate the hazard of accidently spraying paint on another person.

WARNING

Spray-painting will be done only in areas designated for that use. No personnel may enter an area of spray painting without protection until 30 minutes after all painting/cleaning is completed.

c. *Cleaning.*

WARNING

Do not use gasoline for cleaning.

WARNING

Avoid skin contact with cleaning solvents. Wear impervious gloves, eye protection and respirator.

d. *Fire Hazards.*

WARNING

Observe fire regulations when using paints, lacquers, primers, removers and thinners; many are highly flammable. Keep away from heat, flames and sparks.

WARNING

Post "NO SMOKING" signs in and within a 50 foot radius of paint spraying and storage areas.

WARNING

Dried spray-paint dust can pose an extreme fire hazard. Remove and dispose of this dust daily in accordance with AR 420-47. The danger of fire can be materially reduced by the use of a water-wash or waterfall type of spray booth.

WARNING

There will be no open flame or spark-producing equipment (e.g. electric sanders) within a 20 foot radius of any spray-painting area.

WARNING

Wiping cloths soaked with paint residue or oil must either be destroyed or kept in a container meeting NFPA standards to prevent possible spontaneous combustion.

e. Safety Equipment.

WARNING

Personal protective equipment (PPE) must be worn during mixing, painting, cleaning, abrasive blasting, grinding, buffing, or where compressed air is being used. Approved respirators must be worn when required. Refer to para 1-7 for specific PPE and respiratory requirements.

f. Hazardous Materials.

WARNING

Unusable paint mixtures, paint components, primers, thinners, removers, or other materials may be considered hazardous waste and require disposal, in accordance with Federal, State, DOD, and DA hazardous waste regulations. Consult local environmental personnel for proper disposal guidance. This applies to certain dried paint/primer waste.

WARNING

Many materials are carcinogens. Pay careful attention to all warnings, cautions, notes, and safety procedures when using any paints, primers, solvents, and cleaners.

WARNING

Extreme care must be observed in the handling of paints containing mercury or other fungicides to prevent poisoning or skin irritation.

WARNING

Avoid skin contact with thinner. It can cause a skin rash.

CAUTION

Components of different colors or specifications are not interchangeable. Components from different manufacturers may not be mixed.

WARNING

Persons known to be allergic to isocyanates shall not paint with polyurethanes.

g. Temperature.

WARNING

Before welding, soldering or brazing a painted assembly, remove paint finish from assembly.

WARNING

Apply only heat-resisting paints (para 2-34) to items attaining temperatures of more than 400°F (204°C), such as manifolds, exhaust pipes, and mufflers. Other paints, at these temperatures, may produce toxic vapors and/or equipment damage.

CAUTION

Coatings should not be applied at temperatures below 50°F (10°C).

h. *Equipment Hazards.*

WARNING

Only electrical equipment/wiring conforming to NFPA Article 70 will be used where spray-painting is being done.

WARNING

All electrical equipment must be properly grounded before starting any painting procedures.

CAUTION

Mask intake and exhaust parts, breathers, etc., carefully to prevent dust, solution, water, or metal conditioner from entering the engines of vehicles.

CAUTION

Do not use solvents or paints with solvents on electronic equipment as this may cause changes in performance. Refer to Appendix C for painting of electronic equipment.

CAUTION

Do not use petroleum base products on natural rubber as they are destructive agents.

CAUTION

Do not paint/overspray CARC or other solvent-containing paints on vinyl, rubber or lacquer-coated items, as these items are not solvent resistant.

CAUTION

Do not apply CARC to flexible items. Due to its rigidity, the CARC finish may crack when item is bent.

CAUTION

Do not use spray equipment containing any aluminum components to apply coatings formulated with chlorinated solvents (such as MIL-C-46168, Type III).

1-11. OTHER PUBLICATIONS

Additional safety and environmental information is contained in the following publications:

a. AR 40-5, Preventive Medicine.

b. AR 200-1, Environmental Protection and Enhancement.

c. AR 200-2, Environmental Effects of Army Actions.

d. AR 420-47, Solid and Hazardous Waste Management.

e. CFR 1910, OSHA Safety and Health Standards.

f. DODI 4145.19-R.1, Hazardous Materials Handling and Storage Criteria.

g. NFPA Article 70, The National Electric Code.

h. TB MED 502, Occupational and Environmental Health Respiratory Protection Program.

i. TB MED 514, Guidelines for Controlling Health Hazards in Painting Operations.

CHAPTER 2

UNDERCOATS, FINISH MATERIALS, AND RELATED PRODUCTS

Section I. GENERAL

WARNING

Before beginning any painting related activity, read Chapter 1, Section II, Safety Summary.

2-1. APPLICATION

This chapter is intended to serve as a general guide to the selection of suitable materials, procedures, and systems for painting and otherwise finishing metal and wood surfaces. If the correct finish system (see Chapter 3) is used and properly applied, it will keep maintenance to a minimum. Otherwise, moisture or other substances will penetrate the coating and cause the metal to corrode or the wood to rot. Usually the finish coat alone will not provide sufficient protection. For example, lusterless olive-drab enamel, which is somewhat porous, offers relatively little protection; its main function is camouflage. The required protection is provided mainly by primers that, for metal, contain rust-inhibiting pigments, and for wood, have high moisture resistant qualities.

2-2. FINISH SYSTEMS

a. Protective coatings are applied to metal and wood surfaces to protect them from the destructive action of moisture and other injurious agents. In addition, colored coatings improve the appearance of the surfaces to which they are applied and serve to denote the military organization to which the item being painted belongs. Coatings must also resist weathering, cleaning, fumes, oil, the action of fungi, and other causes that impair their protective qualities.

b. Because no single finish material can fulfill all of the requirements mentioned above, finishes, as applied to both metal and wood, are usually composed of two or more materials, each of which serve a definite purpose in the combination coating known as a "finish system." Detailed information on finish systems may be found in chapter 3 and in MIL-STD-171, MIL-STD-709, MIL-T-704, MIL-STD-193, MIL-STD-194, and MIL-F-14072.

2-3. FINISH SYSTEM MATERIALS

a. Fillers. Fillers are heavy-body materials, usually in paste form, that are used to fill depressions and holes and provide a smooth surface after sanding.

b. Primers. Primers are used on metals to provide a corrosion-resistant coating to which the subsequent finish coat will firmly adhere.

c. Sealers. Sealers are used to fill or seal the pores of wood and prevent the contamination of a finish coat by the "bleeding" of an underlying stain or colored filling material. Certain sealers also contain fungicides.

d. Thinners. Tinners make paint workable by adjusting the paint or coating consistency for easy application.

e. Topcoat or Finish Coat. This is the final coat in a finish system. It may be enamel, lacquer, paint, or varnish, depending on the service requirements desired.

Section II. FILLERS

2-4. GENERAL

Fillers, like primers and sealers, are undercoats used to prepare metal or wood surfaces for subsequent and final coats of enamel, lacquer, paint, or varnish. They are heavy-bodied pigment materials, and except for the graduation fillers, are applied with a putty knife, saptula, or other similar tool. They are always used in conjunction with finish coats.

2-5. Sealing Compound, Curing (MIL-S-11031)

a. Characteristics. This sealing compound is a two-part material consisting of a black polysulfide base compound and a catalyst to be mixed according to instructions. This compound and the catalyst are contained in a two-compartment container. After curing, the compound forms a rubber-like material and provides satisfactory adhesion.

b. Use. This compound is used for sealing and plugging exposed holes in fire control instruments, such as holes for setscrews, adjusting screws, and slugs that are accessible from the outside of the instrument.

c. Curing.

WARNING

The catalyst used in this sealer contains a lead compound. Avoid contact with skin. Wash hands after use.

The compound cures in 72 to 96 hours at approximately 80°F (27°C).

2-6. Sealing Compound, Noncuring (MIL-S-11030, Type I, Class I)

a. Characteristics. This is a homogenous, stable, noncorrosive, and nontoxic compound. It is thermoplastic and noncuring, and is not affected by oil or temperatures between -65°F to 180°F (-53.89°C to 82.22°C).

b. Use. It is used for the static sealing of glass to metal instruments and to cover visible headless screws (except adjusting screws). The compound is to be applied prior to painting.

2-7. Filler, Graduation or Engraving (TT-F-325)

a. Characteristics. This is a paste-paint type filler that adheres firmly to the surface to which it is applied. It is issued in black, deep red, white, and translucent white colors. It provides maximum legibility on graduated scales.

b. Use.

(1) For filling in the engraved graduation scales of instruments.

(2) For small-arms sight graduations.

(3) To replace similar material to equipment that has been removed by cleaning operations.

c. Application. Fill the indentations with the paste by brushing, then wipe across the indentations with a cloth or small knife blade. This action will press the paste into the indentations and remove most of the excess paste. Wash the remainder from the surface before it sets with soap and water; rinse with clear water and allow to dry.

d. Drying Time. Air-dry for 12 hours before handling. For the filler to dry hard, air-dry for 24 hours. A finish coat may be applied over the graduated element after the filler has air-dried for 2 hours.

Section III. PRIMERS

2-8. General

Primers are applied to metal to provide an initial coating to which a second coating (i.e. a topcoat) will firmly adhere. The pigment composition of primers for ferrous-base metals usually consists of iron oxide, titanium dioxide, lead chromate, red lead, zinc chromate, zinc dust, zinc oxide and zinc phosphate, or a mixture of these. Zinc chromate is the principal pigment in primers used on aluminum, magnesium, alloys, and on items that will be exposed to damp or wet operating conditions.

2-9. Primer Coating, Epoxy-Polyamide, Chemical and Solvent Resistant (MIL-P-23377, Type I)

a. Characteristics. This primer specification covers two types of two-component, epoxy-polyamide, chemical and solvent resistant primers formulated for spray application and compatible with polyurethane topcoats. They are lead free but contain chromate.

b. Use. They are intended for use of pretreated aluminum alloy surfaces as a corrosion inhibitive, chemical resistant, strippable, epoxy primer that is compatible with urethane topcoats. It should only be used on nonferrous surfaces. It is a primer for use with CARC paints.

c. Application. Thoroughly mix each of the components separately. Component B is slowly poured into Component A with constant stirring until a one-to-one volume ratio is achieved. An induction of 30 minutes is necessary after mixing. Epoxy primers have a pot life of 15 hours.

d. Drying Time. The primer will dry to touch in 30 minutes and dry hard within six hours.

e. Thinner. Thin with MIL-T-81772, Type II. When spraying, thin to a viscosity of about 16 seconds in a viscosimeter cup.

2-10. Primer Coating, Epoxy (MIL-P-52192)

a. Characteristics. This is an air-drying or baking, chemical resistant, epoxy-resin, lead and chromate formulation, compatible for use with polyurethane topcoats.

b. Use. It is intended for used on pretreatment ferrous surfaces. It may be baked at 3000F for 20 minutes, or at lower temperatures for a longer period of time. It has good exterior durability and will withstand 1,1,1 trichlorethane degreasing vapors. It is a primer for use with CARC paints.

c. Application. Thoroughly mix each of the components separately. Mix four parts of Component A to one part of Component B by volume, and stir until well blended. An induction of 30 minutes is necessary after mixing. Epoxy primers have a pot life of 15 hours.

d. Drying Time. The sprayed primer will be set to touch in 10 minutes and dry hard within one and one-half hours.

e. Thinner. Thin with MIL-T-81772, Type II. This primer may be thinned up to 20 percent by volume.

2-11. Primer, Epoxy Coating, Corrosion Inhibiting, Lead and Chromate Free (MIL-P-53022)

a. Characteristics. This is a two-part, flash drying corrosion inhibiting, lead and chromate free epoxy primer for use on pretreated ferrous and non-ferrous metals which must meet air pollution requirements.

b. Use. It may be used to replace MIL-P-23377 and/or MIL-P-52192 where exposure to lead and chromate pigments is not permitted. It is a primer for use with CARC paints.

c. *Application.* Thoroughly mix each of the components separately. Mix four parts of Component A with one part of Component B by volume and stir until well blended. An induction of 30 minutes is necessary after mixing. Epoxy primers have a pot life of 15 hours.

d. *Drying Time.* The sprayed primer will be set to touch within five minutes and dry hard within 90 minutes.

e. *Thinner.* Thin with MIL-T-81772, Type II. This primer may be thinned up to 20 percent by volume.

2-12. Primer Coating, Epoxy Water Reducible, Lead and Chromate Free (MIL-P-53030)

a. *Characteristics.* This primer is a water reducible, air-drying, corrosion inhibiting, two-part epoxy system. It is lead and chromate free.

b. *Use.* It is intended for use on pretreated ferrous and non-ferrous substrates, and is compatible with polyurethane topcoats. It contains no more than 2.81 pounds per gallon (340 grams per liter) of volatile organic compounds (VOC) as applied. It is a primer for use with CARC paints.

c. *Application.* Thoroughly stir Component A by hand until uniform. Mix Component B with Component A in the volume specified by the manufacturer. An induction of 30 minutes is necessary after mixing. Epoxy primers have a pot life of 15 hours.

d. *Drying Time.* The sprayed primer will be set to touch within 45 minutes and dry hard within two hours.

e. *Thinner.* Reduce the admixed primer with water. When spraying, this primer should have a viscosity of 20 seconds in a viscosimeter cup.

2-13. Enamel, Rust-Inhibiting, Olive-Drab (TT-E-485)

a. *Characteristics.* This is a combination air-drying and baking enamel that provides a smooth, semigloss, olive-drab film, possessing excellent corrosion-inhibiting properties.

b. *Use.*

(1) As a one-coat painting system over phosphate-treated or solvent cleaned steel surfaces, such as sheet metal, metal ammunication containers, and gasoline drums.

(2) For a two-coat system consisting of a primer and top coat alkyd finish, for general use.

(3) As a baked primer (in a two-coat semigloss baked finish) on steel and wood.

(4) As an undercoat for all metals except aluminum and magnesium.

c. *Application.*

(1) Brushing. Apply as issued or thin with not more than five percent by volume of thinner.

(2) Spraying. Apply after thinning with not more than 15 percent by volume of thinner.

(3) Roller. Thin to the consistency recommended by the manufacturer.

d. *Thickness of Coating.* This depends on the purpose of the application. When used as a priming coat, a thin (0.4-0.6 mils thickness), uniformerly applied coat is effective. When used as an intermediate or top coat, a fairly heavy coating should be applied (0.9-1.1 mils).

e. Drying Time.

(1) For types I, II, and III, air dry for 16 hours or bake for 45 minutes at 250°F (121.11°C), before handling. For type IV, allow eight hours of air-drying time or bake for 30 minutes at 300°F (148.89°C).

(2) Types I, II, and III require 72 hours of drying time to reach full hardness, or bake for 45 minutes at 250°F (121.11°C) and air dry for 24 hours. For type IV, bake for 30 minutes at 300°F (148.89°C).

f. Thinners. If thinning is required, use mineral spirits paint thinner, TT-T-291, or synthetic enamel thinner, TT-T-306.

WARNING

This enamel contains toxic pigments. Do not breathe the mist from the enamel when spraying. Also, spray dust from this enamel is subject to spontaneous combustion if allowed to accumulate on surfaces of enclosures or booths. Remove spray dust by scraping surfaces at frequent intervals.

Paragraph 2-14 deleted.

2-15. PRIMER COATING (TT-P-636)

a. Characteristics. Primer coating TT-P-636 is a red or brown iron oxide, alkyd-resin base primer that thas good adhesion, durability and flexibility, and covers in one coat. It has good resistance to weathering, although its intended use is as an undercoat.

b. Use.

WARNING

This primer must NOT be used on the inside of drinking water tanks or on amphibious vehicles.

(1) As a priming coat on synthetics, wood, bare or phosphate-treated ferrous metal parts (of motor vehicles, guns, gun mounts, tanks, and metal shipping containers), and other Ordnance materiel.

(2) As a sealing undercoat on the wooden parts of motor vehicles.

CAUTION

This primer should not be used on steel vehicles, vessels, or structures that are exposed to acid fumes.

c. Application.

(1) Brushing. Apply as issued or by thinning with not more than five percent by volume of thinner. Apply one coat of primer for a two-coat finish and two coats for a three-coat finish.

(2) Spraying. For spraying, thin with 15 percent by volume of mineral spirits paint thinner, TT-T-291, or synthetic enamel thinner, TT-T-306.

d. Drying Time.

(1) Before touching, air-dry for 20 minutes to two hours.

(2) For thorough drying, air-dry for 18 hours or bake for 45 minutes at 250°F (121.11°C).

(3) Recoating is permissible after air-drying for 10 hours.

e. Thinner. Thin primer as required with mineral spirits paint thinner, TT-T-291, or gum spirits turpentine, TT-T-801.

NOTE

Olive-drab, rust-inhibiting enamel, TT-E-485, may be used in lieu of enamel TT-P-636.

2-16. PRIMER COATINGS, RUST-INHIBITING (TT-P664)

a. *Characteristics.* This is a fast-drying primer that provides a hard, smooth, satin finish on iron or steel and requires no sanding.

b. *Use.* Use as a primer coat on bare iron or pretreated steel, or when topcoating with an enamel finish coat. This primer may also be used as an intermediate coat over synthetic enamel when nitrocellulose lacquer is applied as a topcoat. It requires no sanding prior to applying the topcat.

c. *Application.* Apply primer by dipping, brushing, or spraying, when thinned as recommended by the manufacturer.

d. *Drying Time.* After 25 minutes the coating is thoroughly dried.

e. *Thinner.* Thin with synthetic enamel thinner, TT-T-306, or by the manufacturer's specified thinner.

2-17. PRIMER, LACQUER, RUST-INHIBITING (MIL-P-11414)

a. *Characteristics.* This is a quick-drying, rust-inhibiting, cellulose nitrate primer. It is pigmented in one type and one grade only. The color is characteristic of red or brown iron oxide pigments.

b. *Use.* Use as a base for lusterless, semigloss, and full gloss lacquer finishes in connection with automotive and general usage. See MIL-STD-171 or the desired type of finish system.

c. *Application.* Apply primer by dipping or spraying when thinned with one part by volume of lacquer thinner, TT-T-266.

d. *Drying Time.* The primer will dry to full hardness in 48 hours, however it will be dry to touch in three minutes.

e. *Thinner.* Thin lacquer with lacquer thinner, TT-T-266.

2-18. PRIMER COATING, ZINC CHROMATE, LOW MOISTURE SENSITIVITY (TT-P-1757)

a. *Characteristics.* This primer has corrosion-inhibiting properties and is prescribed for use on materiel and equipment when severe corrosive conditions exist, such as exposure to salt water or spray.

b. *Use.* Use on iron and steel exposed to acid and salt water spray, on aluminum or magnesium surfaces, for priming of aluminum or magnesium alloys, and on all contact surfaces with other metals or wood.

c. *Application.* Application is by brushing or spraying.

d. *Drying Time.* Air-dry the primer for 15 minutes before handling. It takes 24 hours of air-drying or baking for 45 minutes at 250°F (121.11°C) to dry to full hardness. Recoating may be started after 30 minutes of airdrying.

e. *Thinner.* Thin with toluene, TT-T-548, or synthetic enamel thinner, TT-T-306.

2-19. PRIMER COATING, WATER REDUCIBLE, EPOXY, ESTER-LATEX TYPE, LEAD AND CHROMATE FREE (MIL-P-53032)

a. *Characteristics.* This primer is a water reducible, air-drying, epoxy, ester-latex, resin base system that is lead and chromate free and meets air pollution requirements for solvent emission.

b. *Use.* It is intended for use on pretreated ferrous metals, pretreated aluminum, and wood. It is intended for use on the inside of potable water tanks for marine use, for steel exposed to long-term weathering, or as a lacquer-resistant primer.

c. *Application*. Reduce five parts by volume of primer with up to one part by volume of tap water. Stir well and filter.

d. *Drying Time*. The sprayed primer will set to touch in one hour and will dry hard in 18 hours.

e. *Thinner*. Thin with tap water. Do not exceed 16 percent water by volume.

Section IV. SEALERS

2-20. GENERAL

Sealers are usually unpigmented. They are used in most applications to seal the pores of wood and to serve as an undercoat over which one or more finish coats will be applied. Sealers are also used to prevent the bleeding of underlying substances such as stains, or to prevent resin, from knots, from entering into the finish coat. In some applications where the decorative feature of the finish is of less importance than its sealing and protective qualities, a sealer is used for both primer and final finish coats. Some types contain fungicides. Sealers are not normally used with CARC coatings.

2-21. OIL, LINSEED, RAW (TT-L-215)

a. *Characteristics*. This is a yellowish, transparent vegetable oil extracted from flaxseeds. When exposed to the atmosphere, this oil dries or cures by oxidation to form a tough and flexible film.

b. *Use*. Linseed oil is used as a vehicle for paints, a thinner for paints, and in the formulation of oleoresinous varnishes and enamels. It can also be used as a sealer on bare wood.

c. *Application*. Apply by dipping or brushing. For dipping. allow the wood, which must be completely dry, to soak in the oil for 24 hours before removing from the oil, wiping the excess with a squeegee or cloth, and allowing to dry.

d. *Drying Time*. Air-dry (cure) for about 16 hours. A second coat may then be applied.

WARNING

Wiping-cloths soaked with linseed oil must either be destroyed or spread out to dry in a well-ventilated area to prevent possible spontaneous combustion.

2-22. STAIN, WOOD (TT-S-270)

a. *Characteristics*. This is a penetrating, wipe-off type stain containing a fungicide.

b. *Use*. The stain partially penetrates clean dry wood, and is used where camouflage and a protection against fungi is desired.

c. *Application*. Apply the stain as issued by brushing, dipping, spraying, or wiping.

WARNING

This stain contains a toxic ingredient that is a skin irritant. Wear gloves when handling.

d. Drying Time. Air-dry for 30 minutes before handling; allow 24 hours for drying to full hardness. Recoating may be accomplished after 30 minutes of air-drying.

e. Thinners. If thinning is required, use mineral spirits paint thinner, TT-T-291.

2-23. VARNISH, SHELLAC (TT-S-300)

a. Characteristics. Shellac varnish is a solution or "cut" of a resin made from the secretion of certain insects. It dries to a transparent film that is soluble in shellac thinner. It is not durable under exterior exposure conditions.

b. Use. May be used as a filling or sealing coat on wood, or as an intermediate coat to prevent the bleeding of oil-soluble colors. It may also be used as an intermediate coat over an oil-type sealer.

c. Application. Shellac varnishes are usually applied by brushing. They can be sprayed when thinned properly.

d. Thinner. If thinning is required, use denatured alcohol, grade III.

e. Drying Time. Air-dry for 1 hour before handling; recoat after 2 to 3 hours of air-drying. Varnish dries to full hardness after 24 hours.

Section V. THINNERS

2-24. THINNER, AIRCRAFT COATING (MIL-T-81772)

This thinner is used with CARC topcoats and primers. Type I is used with MIL-C-46168, MIL-C-53039, and those CARC epoxy enamels and primers using MIL-T-81772, Type II, if Type II is not available. Type II is used with MIL-C-22750, MIL-P-53022, and MIL-P-23377, Type I.

WARNING

Wear impenetrable gloves when using this thinner.

2-25. OIL, LINSEED, RAW (TT-L-215)

When used as a thinner in pigmented coating materials, it reduces the pigment-to-oil ratio, resulting in a thinner coat. Paints thinned with linseed oil are used as sealing and priming coats on bare wood.

2-26. THINNER, ENAMEL, SYNTHETIC (TT-T-306)

This thinner is a mixture of volatile coal-tar and petroleum derivatives. It is used to thin synthetic resin-base varnishes enamels. Do not use to thin lacquer.

2-27. THINNERS, LACQUER, CELLULOSE NITRATE (TT-T-266) AND LACQUER (MIL-T-6095)

Lacquer thinner TT-T-266 evaporates much less rapidly than lacquer thinner MIL-T-6095. These thinners are intended for use during periods of high humidity to prevent the condensation of moisture on the surface of the lacquer film. Such moisture seriously impairs the lacquer coating.

2-28. THINNER, PAINT, MINERAL SPIRITS (TT-T-291)

This is a water, white-petroleum derivative similar to and used interchangeably with dry-cleaning solvent. It is used as a thinner for asphalt paints and oleoresinous enamels, paints, and varnishes, except synthetic, resin-base enamels and varnishes. It will curdle or decompose these latter materials. Do no use to thin lacquer.

2-29. TURPENTINE. GUM SPIRITS (TT-T-801)

This is a clear, volatile liquid obtained by distillation of the gum (oleoresin) of living pine trees. It is a good solvent for many resins and is the preferred thinner for oil paints and varnishes with a linseed oil vehicle. Its evaporation rate is relatively slow. Do not use to thin lacquer.

2-30. DELETED

Section VI. TOPCOAT OR FINISH COAT

2-31. CHEMICAL AGENT RESISTANT COATINGS (CARC)

a. General. Chemical Agents pose a devastating threat to sustained readiness in a combat environment. CARC paints were developed to minimize the impact of this threat. CARC paints are relatively impermeable coatings which do not absorb/desorb chemical agents, and which do not break down when decontaminated.

(1) A common misconception is that CARC paints present greater health/safety/environmental hazards than do other paints. In fact, the health and safety requirements for CARC are the same as those for all paints. And, although CARC paints are currently more expensive and require additional care in application, their durability make overall life cycle costs/efforts less than those of other paint systems.

(2) There are currently three CARC paints:

(a) MIL-C-46168 a two-component aliphatic polyurethane used on exterior surfaces and those interior surfaces frequently exposed (eg, ramps, hatches).

(b) MIL-C-53039 a single component aliphatic polyurethane used wherever MIL-C-46168 may be used.

(c) MIL-C-22750 a two-component epoxy polyamide enamel used only on interior surfaces.

b. Coating, Aliphatic Polyurethane, Chemical Agent Resistant (MIL-C-46168) Two-Component.

(1) Characteristics. This specification covers both camouflage and non-camouflage color chemical agent resistant aliphatic polyurethane coatings (CARC). CARC is designed for easy decontamination after liquid chemical agent exposure. It is available in a standard formula (Type II), and a high-solids VOC compliant formula (TYPE IV). Types II and IV are all lead and chromate free.

(2) Use. CARC is intended for use over new or previously painted surfaces. It is applied over pretreated surfaces after priming with an epoxy primer (MIL-P-53022, MIL-P-53030, or M IL-P-23377). CARC can be applied over thoroughly prepared existing CARC surfaces. It cannot be applied over lacquer. MIL-C-46168 is for exterior surfaces and interior surfaces routinely exposed to the outside (i.e., door ramps, hatches, etc.).

(3) Application. Pigments of Component A have a tendency to settle and cake due to the solids content. These solids must be dispersed into a smooth, uniform solution prior to the addition of the catalyst. This can best be accomplished by mechanically agitating or stirring Component A for 30 minutes before mixing. The catalyst, Component B, must be a clear to pale yellow liquid and must be free of crystals. A cloudy, milky, or crystalline gel indicates that the catalyst is contaminated and should not be used. If the container for Component B is swollen, do not open it. Dispose of it as a hazardous waste. Both components should always be measured because accuracy is very important. MIL-C-46168 should be mixed four parts by volume of Component A with one part by volume of Component B. Thinning should not be necessary for brush application. MIL-C-46168 can be thinned for spraying by mixing up to one part by volume of the applicable solvent with four parts by volume of the mixed coating. The applicable solvent for Type II and Type IV is MIL-T-81772, Type I. For adequate camouflage properties, it is necessary to apply the coating to a minimum dry film thickness of .0018 inch (.0046 cm). Under certain temperature and humidity conditions, for more even results, it may be advisable to apply two coats of a minimum thickness of .0009 inches (.0023 cm) each. Component B is water sensitive and caution must be taken to ensure water or high humidity does not come in contact with the coating. Mixed coating must be used within eight hours and cannot be stored. Once opened, component B must be used that day or stored in a sealed dry air/airless container.

(4) Drying Time. Curing time increases with lower temperature or higher humidity. At temperatures of 70 F (21°C) and above, MIL-C-46168 will dry within the specification requirements (set to touch in approximately 15 minutes, dries hard in 90 minutes, dries through in 4 hours, with a complete cure within 7 days). At 60°F (16°C), MIL-C-46168 requires twice as long to cure.

CAUTION

Do not mix components of MIL-C-46168 with MIL-C-53039.

CAUTION

Components of different colors are not interchangeable. Component A of one color may not be used with Component B of another color. Components from different manufacturers may not be mixed.

CAUTION

Do not use CARC on items which are flexible. Because of its rigidity, the finish may crack when item is bent.

CAUTION

Spray lines used with epoxy should not be used with polyurethanes without complete flushing or cleaning with solvents.

NOTE

CARC application requires extremely clean surface preparation. Prior to painting, check cleanliness with the red litmus or water break tests.

c. Coating, Aliphatic Polyurethane, Chemical Agent Resistant (MIL-C-53039) Single-Component.

(1) Characteristics. This specification covers both camouflage and non-camouflage color polyurethane coatings. It is a lead and chromate free, single component CARC with a maximum of 3.5 pounds per gallon of Volatile Organic Compounds (VOC) as packaged.

(2) Use. CARC is intended for use over new, pretreated surfaces. It is applied over pretreated surfaces after priming with an epoxy primer (MIL-P-53022, MIL-P-53030, or MIL-P-23377). CARC can be applied over thoroughly prepared CARC surfaces. MIL-C-53039 is for exterior surfaces and interior surfaces routinely exposed to the outside (i.e. door ramps, hatches, etc.).

(3) Application. Thinning should not be necessary for brush application. It can be thinned for spraying by mixing up to one part by volume of the applicable solvent with four parts by volume of MIL-C-53039. The applicable solvent for all areas is MIL-T-81772, Type I. For adequate camouflage properties, it is necessary to apply the coating to a minimum dry film thickness of .0018 inch (.0046 cm). Under certain temperature and humidity conditions, for more even results, it may be advisable to apply to coats of a minimum thickness of .0009 inches (.0023 cm) each. The coating is very water sensitive and caution must be taken to insure water or high humidity does not come in contact with the coating before it is cured. Once opened, MIL-C-53039 must be used within eight hours, unless stored under a layer of MIL-T-81772, type I, or a blanket of moisture free air or dry nitrogen.

(4) Drying/Curing Time. Curing time increases with lower temperature or higher humidity. At temperatures of 70"F (21°C) and above, MIL-C-53039 will dry within the specification requirements (sets to touch in approximately 15 minutes, dries hard in 90 minutes, dries through in 4 hours, with a complete cure within 7 days). At 60"F (16°C), MIL-C-53039 requires twice as long to cure.

CAUTION

Never mix components of MIL-C-46168 or MIL-C-22750 with MIL-C-53039. They are not compatible.

CAUTION

Do not use CARC on flexible items. Because of CARC's rigidity, doing so may cause cracking of the finish.

CAUTION

Spray lines used for epoxy should not be used with polyurethanes without complete flushing or cleaning with solvents.

NOTE

CARC application requires extremely clean surface preparation. Prior to painting, check cleanliness with the red litmus or water break tests.

d. Coating, Epoxy Polyamide Enamel, Chemical Agent Resistant (MIL-C-22750) - Two-Component.

(1) Characteristics. This specification is for use on the interior surfaces of equipment, vehicles, vans, and shelters. CARC is designed for easy decontamination after liquid chemical agent exposure. Interior surfaces which become exterior surfaces upon opening (ramps, hatches, etc.) should be painted with MIL-C-46168 or MIL-C-53039.

(2) Use. CARC is intended for use over new or previously painted surfaces. It is applied over pretreated surfaces after priming with an epoxy primer (MIL-P-53022, MIL-P-53030, or MIL-P-23337).

(3) Application. Pigments of Component A have a tendency to settle. Stirring for 20 minutes is necessary to disperse these solids into a smooth, uniform solution prior to the addition of a catalyst. The catalyst, Component B, must be clear. Thickening or gelling with the presence of crystals indicates that the catalyst is not usable. M IL-C-22750 should be mixed in accordance with manufacturer's instructions. The mixed components shall stand for an induction time specified by the manufacturer before using. This coating can be thinned, if necessary, by using MIL-T-81772, Type II. For adequate resistance properties, the coating should be applied to a minimum dry film thickness of .0018 inch (.0048 cm). Mixed coating must be used within the pot life specified by the manufacturer, and cannot be stored. Component B is water sensitive and caution must be taken to insure water or high humidity does not come in contact with the coating before it is cured.

(4) Drying Time. At 70"F (21"C) and above, MIL-C-22750 will dry within the specification requirements (sets to touch in approximately 20 minutes, dries hard in 90 minutes, dries through in 4 hours, with a complete cure within 7 days).

CAUTION

Components of different colors are not interchangeable. Component A of one color may not be used with Component B of another color. Components from different manufacturers may not be mixed.

CAUTION

Neither component of MIL-C-22750 is compatible with the single component MIL-C-53039, and should never be mixed with it.

CAUTION

Spray lines used for epoxy should not be used with polyurethanes without complete flushing or cleaning with solvents.

NOTE

CARC application requires extremely clean surfaces. Prior to painting, check cleanliness with the red litmus or water break tests.

e. CARC Shelf Life Extension.

(1) To determine if the shelf life of CARC can be extended, perform the following inspections:

(a) Condition in container - should be no excessive skinning, hard settling or resin separation.

(b) Viscosity - should be no excessive increase in viscosity from specification requirements.

(c) Drying time - should conform to specification.

(d) Application - should conform to specification label instructions.

(e) Thinning - should thin adequately with designated thinners.

(f) Gloss - should conform to specification requirements.

Change 3 2-13

(g) Color - should conform to specification requirements.

(2) If paint meets criteria in (a) through (g), shelf life may be extended by 50%. For example, a paint with a self life of 1 year could be extended six months if it passes inspection above.

(3) For further information about CARC paint inspection, testing and shelf life extension, write to: Commander, U.S. Army Belvoir Research, Development and Engineering Center, ATTN: STRBE-VO, Ft. Belvoir, VA 22060-5606.

2-32. ENAMELS

a. *General.* Enamels are pigmented finishing materials that, in general, dry to a hard gloss, semigloss, or lusterless finish. The nonvolatile vehicles in enamels may be oils, natural or synthetic resins, soluble cottons, or their combinations. For CARC enamels, refer to para 2-31d.

b. *Walkway Compound, Nonslip and Walkway Matting, Nonslip (MIL-W-5044).*

(1) *Characteristics.* This enamel provides a very coarse, gritty coating, similar to coarse sandpaper. It is to be applied over any previously painted or primed surface, including surfaces previously painted with CARC, alkyd, or enamel.

(2) *Use.* A heavy coating is to be applied on surfaces that might become wet in order to provide a more secure footing. For painting tank turret floors, a heavy coating of white enamel is to be used.

(3) *Application.* When brushing, apply as issued or thin to not more than 5 percent by volume. If enamel is to be sprayed, thin to 15 percent by volume. Use thinner specified by the manufacturer. The enamel can also be trowelled on; apply as issued. The thickness of the coating should be from 1/32 to 1/16 of an inch to insure retention, with maximum durability, and nonskid properties. For tactical equipment, apply compound over CARC primer, and apply CARC topcoat cover compound; for non-tactical equipment, do not apply CARC topcoat over the compound.

c. *Enamel, Synthetic, Gloss (T-E-489).*

(1) *Characteristics.* This is a high-gloss, air-drying, alkyd-resin base enamel, with excellent weather-resistant properties. It is flexible and has satisfactory gloss and color retention.

(2) *Use.* This enamel is used on exterior and interior metal surfaces, particularly on smooth exterior surfaces. The main use of the enamel is for refinishing automobiles. For this purpose, it should be noted that when alkyd enamels of this type are applied to steel surfaces, the bare or phosphatized steel should first be given a priming coat with an oxide, zinc-chromate alkyd, or priming surfacer.

(3) *Application.* Apply as issued when brushing on; thin in accordance with the manufacturer's instructions when spraying. The enamel dries hard in 8 hours and dries through in 48 hours. It is ready for recoating after 24 hours of air-drying.

d. *Enamel, Synthetic, Lusterless (7T-E-527).*

(1) *Characteristics.* This is a combination air-drying and baking enamel with an oil-modified, alkyd-resin base. It has satisfactory weather characteristics with regard to chalking, fading, and color changes, but because of film porosity, the enamel is applied in a finish system requiring the use of a corrosion-inhibiting primer.

(2) *Use.* Apply the enamel over a primer when weathering conditions are encountered. Use as a two-coat, lusterless, alkyd finish for both general use and on the outer surfaces of optical instruments.

(3) *Application.* When brushing on, thin to a ratio of not more than 5 parts of thinner to 95 parts of paint. When spraying, thin to a ratio of not more than 15 parts of thinner to 85 parts of paint.

(4) The required thickness for acceptable covering depends to a great extent on the color of the paint. Black, gray, red, and olive-drab colors have satisfactory hiding power, and a relatively thin film will suffice. Orange and yellow paints require a film of at least twice that thickness. White enamel requires a heavy film to obtain complete hiding.

WARNING

The dried spray dust of lusterless enamels is an extreme fire hazard. Remove this dust daily. The danger of fire can be greatly reduced by the use of a water-wash or waterfall type of spray booth.

e. Enamel, Synthetic, Semigloss (TT-E-529).

(1) Characteristics. This is an alkyd-resin base enamel issued in two types: Class A, air-drying, and Class B, baking. Both have satisfactory weathering qualities.

(2) Use. Use on exterior applications when extreme weathering conditions exist.

(3) Application. To brush on, apply as issued or by thinning with not more than five parts by volume of synthetic enamel thinner, TT-T-306, to 95 parts by volume of enamel When spraying, thin with not more than 15 parts by volume of synthetic enamel thinner, TT-T-306, to 85 parts by volume of Class A enamel, or 15 parts by volume of xylene, TT-X-916, to 85 parts by volume of Class B enamel.

(4) Thickness of Coating. A light coat of olive-drab or black enamel will provide sufficient hiding. A considerably heavier coat is required when yellow or white enamel is applied.

(5) Drying Time. Air-dry Class A enamel for eight hours, and bake Class B at 250°F (121.11°C) for 45 minutes before handling.

2-33. LACQUERS

a. General. Lacquers are finish materials that dry by evaporation of the volatile portion only, and deposit a strong film that is generally thinner than the film provided by oleoresinous products. The lacquers described in this paragraph are of this type.

b. Lacquer, Lusterless, Hot Spray (MIL-L-11195).

(1) Characteristics. This lacquer is a one-type and one-grade cellulose-nitrate material which can be applied at room or above-average temperatures.

(2) Use. Use as a one-coat, lusterless finish for projectiles, grenades, etc., and as a two-coat, lusterless finish for automotive and general use.

(3) Application. Apply by dipping when thinned with one (+0.2) part of thinner; hot spray as issued. Apply by cold spray when thinned with one part by volume of thinner to two parts by volume of packaged material. Use the manufacturer's recommended thinner.

c. Lacquer, Automotive, Hot Spray (MIL-L-12277 and MIL-L-52043).

(1) Characteristics. This lacquer is a one-type and one-grade, high gloss, cellulose-nitrate material which can be applied at room or above-average temperatures.

(2) Use. Use as a two-coat or three coat, semigloss finish for automotive materials. Use as a threecoat, semigloss finish on aluminum and magnesium surfaces.

(3) Application. Apply by hot spray as issued; by cold spray, thin with three parts by volume of thinner to five parts by volume of lacquer. Use the manufacturer's recommended thinner.

d. Lacquer, Aircraft (MIL-L-46159).

(1) Characteristics. This is a single-color (Aircraft Green), low reflective, single-component, lead and chromate free lacquer.

(2) Use. Use as a two-coat finish for the exterior surfaces of Army aircraft. It can be used in a wide variety of temperatures (40°F to 90°F (4°C to 320C)) and humidities (15 percent to 70 percent). Excessively high temperature with low humidity can result in dry spray, however, and excessively high temperature with high humidity can result in blushing. Apply over primer MIL-P-23377 or TT-P-1757.

(3) Application. Apply by spray when thinned with up to one and one-half parts by volume of thinner, MILT-19544, to one part by volume of MIL-L-46159. Mix only the amount necessary for use that day. MIL-L-46159 must be thoroughly mixed in its container prior to adding thinner. Constant agitation is also required throughout the painting process. Test on another surface prior to painting aircraft to ensure proper spray viscosity. If excessively high temperatures and/or humidity are encountered, the lacquer/thinner mixture may be combined with up to 30 percent by volume of acrylic lacquer retarder, MIL-E-7125. Apply two coats of mixture to a dry film thickness of 1.5 mils.

WARNING

Use impervious, not cloth, gloves when mixing any component containing MIL-E-7125.

(4) Drying Time. The first coat will air dry in 30 minutes. Lightly wipe prior to applying second coat. The second coat will dry to touch in 30 minutes and is ready for flight in four hours.

2-34. PAINTS

a. General. Paints are mechanical mixtures or dispersions of pigments in a nonvolatile liquid. A volatile solvent or thinner is used to reduce the paint to the proper consistency for application. The pigmented liquid, after application to the surface by brushing, spraying, or dipping, dries to form a solid and opaque coating. An oil paint contains a drying oil or oil varnish as the basic ingredient. A paste paint is one that permits a substantial addition to the vehicle of thinner to obtain the consistency required for application. An asphalt paint contains asphaltum or a similar substance as the principal nonvolatile ingredient; this also provides the coloration of black or brown.

b. Paint, Acid-Proof, Black (MIL-C450).

(1) Characteristics. This paint consists of a material of petroleum or asphalt (bitumen) or a combination of both, thinned with aromatic petroleum naphtha or mineral spirits paint thinner, TT-T-291, to the required nonvolatile content. It contains no drying oils resins, or pigments. This paint is highly resistant to acids.

(2) Use. It is used on metal or wood battery boxes and their supports, and also in the assembly of certain ammunition items.

c. Coating, Underbody (FT-C520).

(1) Characteristics. This compound is a mixture of asphalts, fillers, solvents, and additives processed to meet specified requirements.

(2) Use. Use as a protective underbody coating for automotive equipment and as a sound deadener.

(3) Application. Spray at a temperature of 65°F ±5 F (18.33°C ±2.78°C), with a tank pressure of 80 pounds per square inch (psi) and with no more than 80 psi on the gun nozzle.

d. Paint, Heat-Resisting, Silicone-Aluminum (TT-P-28).

(1) Characteristics. This is a one-type and one-grade, aluminum, heat resistant paint which will withstand solvents and normal weather exposure. It is an air-drying, or air-drying and baking, product with an aluminum pigment of powder or paste, suspended in a liquid vehicle.

(2) Use. It is used for painting equipment where operating temperatures preclude the use of conventional paints and for applications where engraved, stamped, or stenciled lettering would be exposed to temperatures up to 1000°F (537°C).

(3) Application. Spray as issued; brush when required.

e. Paint, Heat-Resisting (For Steel Surfaces) (MIL-P-14105).

(1) Characteristics. This paint contains a blend of ceramic frits, refractories, and pigments in a vehicle of pure or modified silicone resins.

(2) Use. This paint is intended for use on solvent degreased and blasted steel surfaces of components which are subject to temperatures as high as 1,400°F (760°C) and exterior weathering. Components such as mufflers, manifolds, and stacks may be protected by the use of this paint. The paint provides excellent protection against corrosion and chemical attack. It has also been found to perform satisfactorily when applied to parts that do not lend themselves to sand blasting. In application to such parts, however, it is absolutely necessary that all loose rust and tight and loose mill scale be removed by wire brushing and chipping. Coatings should then be applied by brushing, taking care to work the paint well into the roughened surfaces.

(3) Application. Apply by brushing as received, or by spraying reduced in accordance with manufacturer's recommendations.

(4) Drying Time. Paint air dries tack-free within 1 hour. Dries hard when baked at 400°F (204°C) for 30 minutes.

f. Coating, Red Fuming, Nitric, Acid-Resistant (MIL-P-22636 and MIL-P-14458).

(1) Characteristics. This coating consists of a primer conforming to MIL-P-22636 and a paint conforming to MIL-P-14458.

(2) Use. When using red fuming, nitric, acid-resistant coating, refer to the specifications on this material. These are listed in the Reference appendix in the back of this publication.

g. Paint, Stencil, Flat (TT-P-98).

(1) Characteristics. Stencil paint is of one grade and two types. The paint-consistency type comes in black, white, and gray primary and secondary colors; the paste-form type comes in black, white, red. and yellow colors.

(2) Use. For marking bales, crates, fiberboard boxes, ammunition, etc.

(3) Application. Apply with a brush and stencil board, or mask to surfaces of varying textures and absorptive properties.

2-35. VARNISHES

a. *General*. Varnishes are nonpigmented liquids that, when applied as a thin film, dry on exposure to the air and provide a protective coating. Most varnishes are clear or translucent, but certain asphaltic base materials, which are used for protection against moisture or acids and for technical purposes such as electrical insulation, are called varnishes although they are black due to the use of dyes.

b. *Varnishing, Moisture and Fungus Resistant (MIL-V-173)*.

WARNING

This varnish contains a fungicidal ingredient that is harmful if ingested. Observe the manufacturer's precautions in its handling.

(1) *Characteristics*. This is a transparent, phenolic-resin base varnish. The presence of this varnish can be determined by inspection under a "black light", which activates a fluorescent dye. The dye can be added for this purpose prior to the application of the varnish. This varnish has a high dielectric strength.

(2) *Use*. It is used as a surface or finish coat on electrical equipment and components such as coils, circuit wiring, and the chassis of radar and radio assemblies as protection against moisture and fungi.

(3) *Application*. Apply by brush as issued, or after thinning with not more than five percent by volume of thinner. For spraying, apply after thinning with not more than 15 percent by volume of thinner. Dipping consistency depends on the type, shape, and construction of components and assemblies. The degree of thinning required is determined by trial. Use the manufacturer's recommended thinner.

c. *Varnish, Spar, Water-Resisting (TT-V-121)*.

(1) *Characteristics*. This is a durable, waterproof varnish with satisfactory weathering qualities. It is composed of high-grade resins and polymerizing drying oils.

(2) *Use*. Use as issued on exterior surfaces where durability is important and a high gloss is not required. It can also be used as a vehicle for exterior enamels where high resistance to the elements is required.

CHAPTER 3
FINISH SYSTEMS

Section I. CLEANING AND TREATMENT OF SURFACES

WARNING

Before beginning any painting-related activity, read Section II, Safety Summary.

3-1. GENERAL.

This chapter describes finish systems. It covers the material to be used and procedures to be followed in the cleaning, treatment, and painting of equipment to provide protection against rust, corrosion, detection, and/or deterioration. For more information on specific systems not addressed in this chapter, refer to applicable finishing documents such as MILSTD-171, Finishing of Metal and Wood Surfaces; MIL-T-704, Treatment and Painting of Materiel; MIL-STD-186, Protective Finishing Systems for Rockets, Guided Missiles, Support Equipment, and Related Materials; MIL-STD-193, Painting Procedures and Marking for Vehicles, Construction Equipment, and Material Handling Equipment; M IL-STD-194, System for Painting and Finishing Fire-Control Material; MIL-STD-709, Ammunition Color Coding; MIL-F-14072, Finishes for Ground Electronic Equipment; and TT-C-490, Cleaning Methods and Pretreatments of Ferrous Surfaces for Organic Coatings.

3-2. SURFACE PREPARATION

   a. *General.* Surfaces to be painted must be thoroughly cleaned. All rust, corrosion products, oil, grease, moisture, dirt, fouling organisms, loose and blistered paint, deteriorated areas of old paint, and other surface contaminants will be ' removed prior to painting in accordance with MIL-T-704. Surfaces that require removal of loose paint shall be prepared in the following manner:

   (1) The initial step shall be to remove all loose paint by light sand blasting or orbital sanding.

   (2) The edges of good paint surrounding the prepared areas shall be feathered using abrasive sanding disks or stainless steel scouring pads.

   (3) The newly cleaned areas shall be washed with liquid detergent cleaner (e.g. MIL-D-16791, Type 1) and thoroughly rinsed with fresh water until the surface can pass a water break test. The surface shall be allowed to dry completely; pretreatment should begin within four hours after cleaning.

   b. *Paint Remover.* Paint and varnish remover will conform to TT-R-251, Type IV, Class A, low viscosity, for horizontal surfaces, and Class B, high viscosity, to be used for vertical and near vertical surfaces. These paint removers will have minimal effect on CARC coated surfaces, however. For CARC-coated surfaces, use paint remover, epoxy, polysulfide, and polyurethane systems, MIL-R-81294, Type I, according to manufacturer's instructions.

   c. *Solvent Cleaning.* Surfaces intended for conventional paint will be cleaned with the currently approved cleaning agent. Surfaces intended for CARC coatings will be cleaned with a solvent conforming to MIL-T-81772 or TT-T-266. Metal surfaces intended for vinyl paints will be cleaned with naptha solvent, Type A, of MIL1-N-15178. Surfaces already painted with vinyl, acrylic, or acrylic nitrocellulose paint will be cleaned with either Toluene, TT-T-548, or Xylene, TT-X-916, Grade A, prior to over-painting. Mineral-spirit type solvents will not be used on surfaces to be coated with paint because these solvents leave an oily film which interferes with proper adhesion.

WARNING

Toluene, xylene, and naphtha are highly flammable. Exercise extreme care when using these solvents. Do not expose to heat or open flames.

3-3. CLEANING OF SPECIFIC SURFACES

a. *General.* Unless otherwise stated in the end item specification, cleaning shall be accomplished by chemical methods (such as solvent cleaning, alkaline cleaning, acid cleaning, pickling, descaling with hydride or paint stripping), by electromechanical cleaning methods (such as electropolishing, electrolyte alkaline, or electrolytic pickling), or by mechanical means such as blasting, chipping, wire brushing, or grinding. After cleaning, all surfaces shall be kept free from dirt, dust, finger marks, and other contaminants. Various surfaces, such as ferrous metals, zinc, aluminum and aluminum alloy, magnesium alloy, wood, and previously painted surfaces, require special handling.

b. *Ferrous Metal Surfaces.* Unless otherwise specified, ferrous metal surfaces to be painted shall be blast cleaned in accordance with Steel Structural Painting Council (SSPC) Specification SSPC-6 to remove milliscale, products of corrosion, dirt, casting, sand, slag, and other foreign substances. Also, when stated, blast-cleaning shall be in accordance with specifications SSPC-5 or SSPC-10, as required (see Steel Structures Painting Council Manual, Volume 2, SSPC-SP6-63 for more information). Blast-cleaned surfaces shall be given a prime coat as soon as possible after cleaning and in no case more than four hours later. Blast-cleaning shall not be used on surfaces which could be damaged, such as machine parts and sheet metal thinner than 0.0625 inch (16 gage U.S. Standard).

(1) Blast-cleaning is optional on components painted for protection during limited storage, from which the paint will be worn off as soon as the equipment is placed in use. Examples are truck assemblies, track roller assemblies (including mounting frames), interiors of weld-type box sections, bulldozer components (including rippers, scarifiers, ejectors, push plates, blades, bowls, and buckets), scrapers and crane shovels, interiors of cement mixer drums, and interiors of aggregate driers. However, these surfaces shall be dry and free from oil, grease, dirt, and rust prior to painting.

(2) Ferrous metal surfaces of vehicles shall be cleaned for painting in accordance with paragraph 3-3b above, except as specified herein. Surfaces of malleable iron or steel castings shall be cleaned as stated in paragraph 3-3b(1) above. Other surfaces that cannot be cleaned by blasting may be cleaned to base metal by chipping, powered wire brushing, or grinding to the required degree specified above for commercial sand blasting. Sheet metal and sheet metal parts of eight gage and thinner may be cleaned to bare metal by acid pickling in accordance with TT-C-490, with a maximum of five percent sulfuric acid included. Old paint may be removed from vehicles requiring repainting by the use of a paint remover.

c. *Zinc Surfaces.* Zinc surfaces, including zinc-coated ferrous material, shall be thoroughly cleaned, as specified in paragraph 3-3a above, to remove all traces of oil, grease, dirt, and other foreign substances.

d. *Aluminum and Aluminum-Alloy Surfaces.* These surfaces shall be thoroughly cleaned, as specified in paragraph 3-3a above, to remove all traces of oil, grease, dirt, and other foreign substances. This shall be followed by a three to five minute immersion or pressure spray in a hot, 10 percent solution of chromic acid, after which the surfaces shall be thoroughly rinsed with clean, warm water to remove excess chromic acid from cavities, joints, and recesses. The concentration of chromic acid shall be checked at regular intervals to ensure that the solution is maintained at the specific strength. Aluminum surfaces that cannot be immersed or sprayed with chromic acid shall be mechanically cleaned, swabbed with a solution of MIL-C-5541, Type II, Grade B, Class 2.

e. *Magnesium Alloy Surfaces.* Magnesium alloy surfaces shall be cleaned in accordance with MIL-M-3171.

f. *Wood Surfaces.* Wood surfaces to be painted shall be dry and cleaned of all dirt, oil, grease, and other foreign substances with a straight, petroleum-aliphatic solvent.

g. *Previously Painted Surfaces.* Any coating showing corrosion, cracking, blistering, or flaking must be sanded down to bare substrate and solvent cleaned. Consider such surfaces bare and treat as required.

(1) *For CARC application.*

Over CARC. CARC may be applied over sound CARC surfaces that have been solvent cleaned. Items painted I with CARC will not normally require stripping. Exceptions are corroded areas and aircraft with severe weight restrictions. These surfaces shall be cleaned of paint by Plastic Media Blasting (PMB) at 40 psi, whenever possible. This is the preferred method of removing CARC primers and enamels. After paint removal, the entire surface will be cleaned in accordance with MIL-STD-186 or MIL-T-704 and tested for cleanliness using the water break or red litmus test.

(2) For application of other finish systems. Previously painted surfaces that are to be painted with finish systems other than CARC should follow the requirements and procedures of the individual finish systems and specifications involved.

3-4. SURFACE TREATMENT

a. General. Bare metal surfaces to be painted with CARC coatings or vinyl paints will be coated immediately after cleaning with pretreatment primer conforming to DOD-P-15328; this pretreatment will be used under conventional paints only where specified. This wash primer will not stick to steel surfaces which have been treated with metal conditioner, MIL-C-1 0578, Type II. Anti-fouling paint will not be applied over bare metal, since the copper in this paint will corrode the steel. Aluminum may require pretreatment with MIL-C-5541. Whenever possible, MIL-STD-193, MIL-T-704, or other applicable finishing documents should be consulted.

b. Use. Pretreatment primer DOD-P-15328, is used as a bonding agent and to provide temporary protection against corrosion. In general, it may be applied to all bare surfaces, both exterior and interior. Ventilate and take proper precautions pertaining to flammable materials.

3-5. TREATMENT OF SPECIFIC SURFACES

a. Treatment. Pretreatment of surfaces is generally used as bonding agent between the surface of the equipment and follow-on coatings. It provides temporary protection against corrosion.

b. Ferrous Metal, Zinc, or Cadmium Surfaces. Ferrous metal, zinc, or cadmium surfaces shall be treated as soon as possible after cleaning, as specified in paragraph 3-3 above, and as follows:

(1) With an organic pretreatment primer conforming to DOD-P-15328, or with a zinc phosphate (Type I) or iron phosphate (Type II) chemical conversion containing in accordance with TT-C-490.

(2) Any evidence of rust or contamination on a previously cleaned surface shall be cause for recleaning prior to painting.

c. Aluminum Surfaces. Aluminum surfaces shall be treated as soon as possible after cleaning, as specified in paragraph 3-3 above, and as follows:

(1) With an organic pretreatment primer conforming to DOD-P-15328, or in accordance with MIL-A-8625 or MIL-C-5541.

(2) Any evidence of corrosion or contamination or previously cleaned surface shall be cause for redeaning prior to painting.

d. Magnesium Alloy Surfaces. Prior to painting, magnesium alloy surfaces shall be treated in accordance with MIL-M-3171, Type I or III. Treated surfaces that become scratched in handling shall be touched up in accordance with MIL-M-3171, Type I.

e. Wood Surfaces. Properly seasoned wood shall be sealed prior to application of CARC with a polyurethane sealer covered by NSN in Table B-12. Single and two component sealers are listed. Glue used during construction with wood shall be treated with sealer after construction. Unless otherwise specified, wood shall be treated prior to sealing in accordance with MIL-T-704; i.e., dried to a moisture content no greater than 20% and pressure treated in accordance with American Wood Preservers Bureau (AWPB) LP-2 (above ground) or LP-22 (ground contact). Only Ammoniacal Copper Arsenate (ACA) or Chromated Copper Arsenate (CCA) preservatives shall be used. Alternate processes are available when repainting or when pressure treatment is not available.

f. Hardware and Hardware Items. Hardware and hardware items such as bolts, capscrews, washers, pins, springs, and grease fittings are not to be cleaned and treated prior to assembly and painting if there is no evidence of rust or corrosion.

g. Corrosion-Resisting Steel Surfaces. Corrosion-resisting steel surfaces shall be cleaned as specified in paragraph 3-3 above, then treated as follows (unless the corrosion-resisting steel has already been passivated and has not been contaminated or depassivated by working, forming, or shaping the end item). The process specified below is primarily a passivating treatment for corrosion-resisting steels and is not cleaning treatment.

(1) Degrease, as in MIL-S-5002.

(2) Immerse for 30 minutes in a solution containing 20 percent by volume of nitric acid and two percent by weight of sodium dichromate at 120°F to 1300F.

(3) Rinse in clean hot water.

(4) Immerse for one hour in a solution containing five percent by weight of sodium dichromate, at 140°F to 160°F.

(5) Rinse in clean hot water.

(6) Rinse in hot water (160°F to 210°F) with the rinse maintained at pH 3 to 5 by the addition of flake chromic acid or proprietary mixtures of chromic and phosphoric acid. Surfaces to be painted shall be treated with wash primer conforming to DOD-P-15328 or MIL-C-8514.

Section II. PAINTING

3-6. APPLICATION

The first coat of paint of primer shall be applied to a dry, clean surface as soon as is practical. The coatings shall be applied in an ambient temperature 50°F, or higher. The paint and surface shall be approximately the same temperature except when hot spray is sued. Painting shall conform to the finish systems listed in tables 3-3 and 3-4 and shall be applied by any method (dip, flowcoat, brush or spray) which will deposit the dry film coat-thickness specified in table 3-1. Panels or subassemblies prepainted prior to the final assembly shall be treated and painted as specified herein. A smooth, even surface, free from runs, sags, or other defects which might interfere with the application and adhesion of subsequent coats, shall be applied. When applying the priming coat, sufficient time must be allowed for the paint to dry prior to applying the finish coat. Baked finishes, except on materials that would be adversely affected by such treatment, will be permitted if the baked finish conforms to performance requirements of the applicable paint specification.

NOTE

CARC primers and coatings cannot be applied using dip or flowcoat methods.

3-7. FILM THICKNESS

The dry film thickness of some of the more common coatings are listed in table 3-1. For other coatings, reference individual specifications.

Table 3-1. Dry Film Thickness of Each Applied Coat

Specification	Pretreatment Coat	Prime Coat	Intermediate Coat	Finish Coat[1]
	Mils	Mils	Mils	Mils
DOD-P-15328	0.3-0.5	-	-	-
MIL-C-22750	-	-	-	min. 1.0[3]
MIL-C-46168	-	-	-	min. 1.8[3]
MIL-C-53039	-	-	-	min. 1.8[3]
MIL-L-52043	-	-	-	0.8-1.2
MIL-P-14105	-	-	-	1.5-2.5
MIL-P-15931	-	-	-	1.8-2.2
MIL-P-23377	-	0.6-0.9[2]	-	-
MIL-P-24441	-	2.8-3.2	2.8-3.2	2.8-3.2
MIL-P-53022	-	1.0-1.5[5]	-	-
MIL-P-53030	-	1.0-1.5[5]	-	-
TT-E-485	-	1.0-1.5	-	1.0-1.5
TT-E-489	-	-	-	0.8-1.2
TT-E-522	-	-	-	1.0-1.5
TT-E-527	-	-	-	0.8-1.2
TT-E-529	-	-	-	0.8-1.2
TT-E-1593	-	-	-	1.3-1.7
TT-P-636	-	1.0-1.5	-	-
TT-P-645	-	1.3-1.7	-	-
TT-P-664	-	1.0-1.5	-	-
TT-P-1757	-	0.7-1.0	-	-

[1]For Army use on exterior surfaces, only forest green lusterless finish coats shall be used, except green 383 shall be the base color when chemical agent resistance is required. Interior surfaces shall be painted as specified. This does not apply to camouflage coatings, which shall be painted as specified on applicable patterns.

[2]For use on nonferrous metals when a chemical agent resistant topcoat will be applied.

[3]For use when chemical agent resistance is required.

[4]For use on ferrous metals when a chemical agent resistant topcoat will be applied.

[5]For use on ferrous and nonferrous metals when a chemical agent resistant topcoat will be applied.

Change 3 3-5

3-8. TECHNIQUES OF MIXING AND THINNING

a. *Method*. The best, quickest, and easiest method of painting is by spraying. Paint rollers are used on large surfaces when spraying is impractical. Paints are brushed on when other methods are impractical or special equipment is not available. In general, the use of brushes is confined to touchup jobs.

b. *Readiness*. In most cases, paints issued ready mixed, hence color blending is not required. CARC paints MIL-C-56168 and MIL-C-22750, however, are issued in a two-component form and require accurate mixing techniques.

c. *Preservation*.

(1) *String*. Stir paints well before use. If the vehicle (liquid portion) has separated from the pigment, pour off most of the liquid portion into a clean container. Stir the thick settled portion (pigment) in the bottom until all chunks are softened and dissolved. Restore the poured off portion a little at a time, stirring constantly with a lifting and beating motion. "Box" the paint thoroughly by pouring it from one container to another several times, stirring the paint for a few minutes between each transfer.

NOTE

Do not "box" lacquer, as this will cause a loss of the liquid portion by evaporation.

(2) *Straining*. When paint stands over a period of time, a skin may form over the surface and the pigment may form into chunks to the extent that stirring will not mix all of the ingredients properly. In such cases, strain the paint through a strainer into a clean container, discarding the residue left ion the strainer. Do not strain CARC coatings, however. CARC coatings which cannot be properly mixed will be resealed and disposed of as hazardous wastes.

(3) *Thinning*.

(a) When it is necessary to thin paint, use a small amount of the prescribe thinner. Because of its volatility, thinner will evaporate from the paint film, leaving practically the same ratio of vehicle to pigment per square foot of surface as the paint would have provided before thinning. The warmer and drier the weather, the less thinner is needed because heat tends to thin the vehicle. More thinner is required in cold weather to hasten the drying and hardening of the film. Thinner should be used with care, as the less used, the more durable the applied coat will be.

(b) Polyurethane coatings may be thinned up to 20 percent by volume with thinner MIL-T-81772, Type I, or with the manufacturer's recommended thinner. Epoxy primers which are admixed (blended) four to one by volume, such as MIL-P-53022, may be thinned up to 20 percent by volume with epoxy thinner MIL-T-81772, Type II. Epoxy enamels and primer which are admixed one to one by volume, such as MIL-C-22750 and MIL-P-23377, usually have satisfactory spray viscosities, but may be thinned with small amounts of MIL-T-81772, Type II, if necessary. Water is used to thin MIL-P-53030.

(c) Paints which contain a slow-drying vehicle may require additional thinner.

(d) Varnish should not be thinned except when used as a primer coat; it should then be thinned with a small amount of its recommended thinner.

NOTE

Do not shake varnish. This may entrap air which will be difficult to eliminate from the film.

(e) To thin synthetic enamels, use synthetic enamel thinner, TT-T-306.

(f) When painting with acrylic or acrylic nitrocellulose lacquers, it is advisable to add 15 to 30 percent by volume of acrylic lacquer retarder, MIL-E-7125. The retarder tends to slow the drying of the lacquer and evens out its viscosity. The amount of retarder will vary depending on temperature and/or humidity.

(g) Should linseed or other oils be used, the ratio of pigment to vehicle should be reduced, giving the paint less hiding power and greater penetrating power. On certain primary coats on wood or plaster, this is desirable.

d. Shop Atmospheric Conditions. If painting is to be done in an enclosed area, efforts should be made to control the temperature to approximately 75°F to 80°F (24°C to 27°C), and the relative humidity to approximately 45 to 50 percent. Humidity may be lowered by raising the shop temperature.

3-9. OPACITY AND COVERING DATA

a. Oil paints. Table 3-2 indicates the approximate area, in square feet, which can normally be covered per gallon of oil paint. This information can be used as a guide in estimating the amount of paint required for a specific job.

Table 3-2. Oil Paint Coverage

Material	Prime Coat (In Sq. Feet)	Second Coat (In Sq. Feet)	Third Coat (In Sq. Feet)
Steel:			
Sheet	400-600	500-600	600-700
Heavy construction	400-550	450-600	550-650
Medium	350-500	450-550	500-600
Light	300-500	400-550	450-550
Wood:			
New	500-600[1]	500-600[2]	500-600
Weathered	300-400[1]	400-500[2]	400-500
Repainted	400-500	500-600	500-600
Concrete and brick	150-300[3]	300-400[4]	350-450
Plaster, etc.	250-350[2]	300-400	400-500

[1] To each gallon of paint is added approximately two quarts of raw linseed oil and one pint of thinner.
[2] To each gallon of paint is added approximately one pint of raw linseed oil and one-half pint of thinner.
[3] Special primer or reducer added.
[4] Approximately one pint of reducer is added to each gallon of paint.

b. Enamels, Varnishes, Lacquers, and Stains. In general, the approximate area, in square feet, which can be covered per gallon, depending on surface and consistency of paint, is as follows: enamels. 400-600: varnishes, 500-700; lacquers, 75-200; and stains, 500-600.

3-10. STORAGE OF PAINT MATERIALS

WARNING

Post "NO SMOKING" signs in paint warehouses.

a. The materials covered in this paragraph include the primers, fillers, paints, varnishes, lacquers, and other liquid products that are required for protective finishes.

b. Store these materials where they will be protected from the elements and extreme temperature changes. While freezing temperatures may cause a separation of some ingredients, which are difficult to mix again with uniform consistency, the majority of the products described in this manual are not damaged by freezing. CARC coatings (MIL-C-46168, MIL-C-53039, and MIL-C-22750) cannot be used after being frozen, however. Low temperatures do tend to increase the viscosity of paints, varnishes, and like materials. This makes their application difficult, and impairs the adhesion if they are applied when temperatures are very low. High temperatures may cause a soapy, foamy condition, or a chemical change of the oils in a paint or varnish, and make them unusable.

NOTE
CARC paints have a shelf life of one year. This one-year shelf life can only be reached if CARC paint is stored at a proper temperature range of between 32°F and 120°F.

c. Up-end containers every 90 days when they are stored on end, or rotate them one-half turn every 90 days when they are stored horizontally.

d. Do not store partially filled containers without tightly installing lids, covers, or caps.

e. Do not store paints, varnishes, or other flammable materials near steam pipes, open flames, or where there is a danger of flying sparks, such as from welding equipment.

f. Paint and paint thinners shall be stored separately from other materials such as grease, oil, and spare parts. Rags, wood, and similar matter shall not be stored in the same area as paints and paint thinners.

g. To avoid possible leakage from rusted containers, protect containers against rain, snow, steam leaks, and other sources of water.

h. Each container should be labeled with complete instructions as to the type of material, the thinning ratio, the thinning material, and color, gloss, and application data. Each container should also be labeled with safety warnings and cautions.

i. Maintain a perpetual inventory of all materials when the volume is large enough to warrant the effort. Install a system of dating for each shipment received. Use the oldest stock first since aging causes certain types of coatings to lose their gloss and to thicken to such an extent that they are rendered useless. Black enamels have a particular tendency to lose their gloss and drying properties upon aging.

j. After the shelf life of a paint has been reached, if samples of the paint conform to the specification requirements for viscosity, drying time, application, thinning, gloss, and color, and if the condition in the container reveals no excessive skinning, hard settling, or resin separation, the shelf life may be extended by 50 percent (i.e. a one-year shelf life would be extended by six months). This includes storage extension for CARC paints.

3-11. TIPS ON PAINTING

Certain basic precautions are applicable to paint, varnish, enamel, and lacquer. The following should be observed at all times:

a. Do not paint over an unclean surface. Be sure all dirt, rust, scale, etc., are removed.

b. Do not fail to stir paint thoroughly.

c. Do not mix one paint with another unless instructed to do so.

d. Do not fail to follow instructions which appears on containers, particularly those concerning safety, the addition of thinner, and the application instructions.

e. Do not apply paint or varnish unless the drying conditions are satisfactory.

f. Do not paint in wet or extremely cold weather (below 50°F (10°C)).

g. Do not apply abnormally heavy coats.

h. Do not add too much thinner.

i. Do not use paint buckets, cans, paint rollers, spray guns, or brushes with are not clean.

j. Do not apply cold paints on varnishes.

k. Do not leave old paint and oil-soaked cloths laying around in the paint shop; they are a fire hazard.

l. Do not fail to clean brushes, paint rollers, and spray guns immediately after using.

m. Do not smoke when painting. Do not smoke near paint storage areas or paint booths.

n. Do not release the tops of pressure-feed material containers before releasing the air pressure.

o. Do not use electrical connections that show any inclination to becoming loose.

p. Do not pour paint out of a container in a manner that obscures the label.

q. Do not fail to strain paint before using, if required. CARC paints cannot be strained, however.

r. Do not fail to remove all traces of wax from surfaces where paint or varnish is to be used.

s. Do not paint without proper respiratory equipment and ventilation.

t. Do not waste paint by spraying behind the item being coated.

u. Do not paint over a moist or wet surface.

v. Do not paint between the ground strap and hull of tanks.

w. Do not paint on operator-instruction plates.

3-12. TIPS ON PAINTING WITH CARC

The following precautions should be observed, in addition to those listed in paragraph 3-11 above, when applying Chemical Agent Resistant Coatings (CARC):

a. Spray lines for epoxy applications should not be used with polyurethane coatings without complete flushing or cleaning with solvents.

b. Test for cleanliness when applying CARC with the red litmus or water break test.

c. Remember to notify the local safety office and preventive medicine support activity prior to initial CARC painting. This also applies to all spray painting operations, regardless of the material used.

d. Do not use CARC for items like manifolds and mufflers that exceed 400°F. Do not use CARC on rubber, lacquer coatings, or vinyl.

e. Use impervious, not cloth, gloves when applying CARC.

f. Do not apply CARC to flexible items. Because of its rigidity the finish may crack when bent

g. When using CARC, mix only the amount needed to do the job (i.e. don't open a large container for a small job) because unused CARC must be disposed of and cannot be stored.

h. Epoxy-polyamide coatings build up thickness quickly. Thick films are detrimental for good adhesion. Do not apply CARC beyond its thickness tolerances.

Section III. SPECIFIC FINISH SYSTEMS

3-13. GENERAL

a. Specific finish systems for both camouflage and non-camouflage materials are addressed in this section and in paragraphs 3-1 through 3-5. Refer to those paragraphs for additional details. Tables 3-3 and 3-4 lists the different finish systems (i.e. combinations of primers, topcoats, and pretreatment materials used on specific types of surfaces) that are most frequently encountered. Details on each of the components covered can be found in Chapter 2 and in the individual specifications. This paragraph and paragraphs 3-14 through 3-16 give additional information on particular finish systems, including those using CARC.

b. Specifications for CARC camouflage colors contain requirements to protect the military equipment against visual and infrared detection and chemical agent contamination. CARC paints have this protective ability and also the ability to be easily decontaminated. Camouflage coatings, especially aircraft coatings, are rough and difficult to clean under field conditions. Cleaning agent, MIL-C-85570, is very useful in cleaning aircraft and other camouflage (alkyd or CARC) painted equipment.

c. For further camouflage paint (QPL) information, see Chapter 4, or contact: Commander, Belvoir Research, Development and Engineering Center, ATTN: STRBE-VO, Fort Belvoir, Virginia 22060-5606.

Table 3-3. Camouflage Finish Systems

Top Coat[3]	Primer	
	Aluminum Surfaces	Steel Surfaces
	TT-P-1757	TT-P-636 TT-E-485[1]
	TT-P-1757	TT-P-664 TT-E-485[2]
	TT-P-1 757 M IL-P-23377	TT-P-664
	TT-P-1757	TT-P-664
MIL-C-53039 MIL-P-53030	MIL-C-46168 MIL-P-53022 MIL-P-53030	MIL-P-23377 MIL-P-53022

[1]When using T--E-485, types II or IV can be used.
[2]When using TT-E-485, type IV should be used.
[3]Where lead and chromate free topcoats are required, use type II of the camouflage specifications, where applicable.

Table 3-4. Non-Camouflage Finish Systems

Type	Surface	Pretreatment	Primer	Coats	Intermediate	Coats	Finish[11]	Coats[1]
A	Metal	DOD-P-15328[4]	TT-E-485 TT-P-636	1	-	-	TT-E-485	1
	Metal (For lead and chromate free)	DOD-P-15328[4]		1	-	-	TT-E-529[1,2,3]	1
	Metal (For marine environment)	DOD-P-15328	TT-P-645	1	-	-	TT-E-485 TT-E-1593	2
	Wood (For lead and chromate free)	See MIL-T-704 See MIL-T-704[10]	TT-P-636	1	-	-	TT-E-529 TT-E-529[1,2,3]	1 1
B	Metal (B-1[13])	-	MIL-P-24441, formula 150	1	MIL-P-24441, formula 150	1	MIL-P-24441, formula as specified	1
	(B-2[14])	-	MIL-P-24441, formula 150	1	MIL-P-24441, formula 150	2	MIL-P-1 5931	2
	Wood	See MIL-T-704	TT-P-636 1	1	TT-P-636	1	TT-E-522	1
C	Metal	DOD-P-15328[4]		-	-	-	TT-E-485[3,9]	1
D	Metal	DOD-P-15328[4]	TT-P-1757[6]	1	-	-	TT-E-529[1]	2
E	Metal	DOD-P-15328[4]	TT-P-1757[6]	1	-	-	MIL-L-52043	2
F	Ferrous metal	DOD-P-15328[4]	MIL-P-53022[12] MIL-P-53030[12]	1	-	-	MIL-C-22750[8] MIL-C-46168[8] MIL-C-53039[8]	2
G	Non-ferrous metal	DOD-P-15328[4]	MIL-P-23777[12] MIL-P-53022[12] MIL-P-53030[12]	1		-	MIL-C-22750[8] MIL-C-46168[8] MIL-C-53039[8]	2

[1]For vehicles or equipment specified in MIL-T-704, paragraph 3.3.4.5, coat shall conform to TT-E-489

[2]A maximum of five percent of chrome yellow, TT-P-381, color 7B, shall be added to enamel, TT-E-485, when it is used as a primer, or the black may be omitted and the resulting yellow enamel used as the primer so as to provide a color contrast between coats. The yellow enamel with the black omitted shall be used as the primer when the color of the finish is other than forest green or dark green. Gloss finish coats shall conform to TT-E-489.

[3]When paint conforming to TT-E-485 is used, the type shall be suited to the method of application.

[4]Alternates specified in MIL-T-704, paragraph 3.2, can be used interchangeably with DOD-P-15328.

[5]TT-E-485 or equal commercial primer.

[6]Color optional.

[7]Camouflage coatings must conform to MIL-T-704, paragraph 3.3.4.1 ,except that camouflage coatings applied over type C finishes must also be compatible with the under coatings used.

[8]For use where chemical agent resistance is required, or for greater durability.

[9]When using this finish and requiring the camouflage forest green color, an anticorrosive primer must be used.

[10]When lacquer resistance is required, TT-E-485, type IV, or TT-P-664 can be used. When camouflage painting is required, the camouflage and coatings shall replace these specified topcoats (See table 3-3).

[12]Parts and components may first be primed with alkyd base primer, which has been allowed to cure for a minimum of 15 days, and then cleaned and primed as specified herein either prior to or at final end item assembly painting.

[13]For general use for water immersion and salt air exposure.

[14]For continuous use in sea water; not for salt air exposure.

3-14. TOUCHUP AND RECOATING.

a. When touching up damaged areas, the procedure should be as similar to the original method of finishing as possible; a clean surface is imperative. If the old finish is in good condition, clean the surface with a compatible cleaning solvent and apply the topcoat. Where general disintegration of the surface is evident, or the under surface is corroded, the coating must be stripped clean from the part. Corrosion must be removed or neutralized by mechanical or chemical treatment, or both, and the surface metal must be penetrated, primed, and then topcoated.

b. Camouflage CARC coatings can be applied on MIL-P-53022, MIL-P-53030, MIL-P-23377, MIL-C-46168, MIL-C-53039, MIL-C-22750, and fully cured alkyd surfaces. All of these surfaces must be clean and free of all contaminants such as oil, grease, fuel, hydraulic/transmission fluid, wax, carbon deposits, sanding debris, water, and fingerprints. Clean surfaces should be tested using the red litmus or water break tests. CARC cannot be applied over lacquer. All lacquer painted items must be stripped down to the epoxy prior to applicable of CARC.

c. Camouflage alkyd coatings can be applied over enamel primed substrates such as MIL-P-52977, MIL-P-52999t, TT-P-636, TT-P-664, TT-P-1757, TT-E-485, or enamel topcoats as long as the surface is clean, dry, and fully cured. Alkyd paints cannot be used over previously CARC painted surfaces.

d. The primary method for determining whether equipment is currently painted with CARC or alkyd is to examine the area near the equipment data plate. The word "CARC" or "ALKYD" should be stenciled nearby. For equipment without a data plate, thoroughly wet a rag with acetone (i.e. fingernail polish remover) and briskly rub the painted surface for 20 seconds. Evidence of actual paint removal onto the rag indicates an alkyd painted surface.

3-15. TARGET MATERIAL.

Target materiel is generally governed by the policy that wooden parts destroyed by bullets will not be painted. Timber frame-supports of sliding targets are, however, given one coat of commercial red paint. The pulleys, sash cord, sash cord clamps, roller brackets, rollers, slide racks, slide irons, and hook bolts of sliding targets are not painted. All parts of the car and track of rolling targets for machine guns, and all parts of sled targets, except snatch blocks, ropes, staves, and pasteboard targets, receive one coat of commercial red paint.

3-16. ENGINE, ENGINE ACCESSORIES, ENGINE COMPARTMENTS.

a. General Instructions.

CAUTION

Mask intake and exhaust ports, breathers, etc., carefully to prevent dust, solution, water, or metal conditioner from entering the engine.

(1) CARC should be used on all surfaces, interior and exterior, of tactical (combat, combat support and ground support) equipment where the temperature does not exceed 400"F. This would include engine compartments, for example. The type and color for interior should be specified to facilitate maintenance or human factors engineering considerations.

(2) Engines, engine components, and powertrain assemblies which are normally painted should be painted consistent with (1) above. For areas that exceed 400"F, paint should conform to MIL-P-14105 or TT-P-28, as applicable. Generally, this means that exposed engines (e.g., on a roadgrader) or engine components should be painted with MIL-P-14105 in a camouflage pattern compatible color. Otherwise TT-P-28 or the manufacturer's coating would be acceptable, because there is no high-temperature CARC.

(3) While the intended use of MIL-P-14105 is on ferrous substrates, it will work on aluminum, but the performance limit of the coating ($1400^{\circ}F$) is well above the melting point of aluminum. TT-P-28 has a performance limit of 1200"F. Neither coating should be used with a primer.

b. Combat Vehicle Air-Cooled Engines. On combat vehicle air-cooled engines, ferrous metal parts are painted with olive-drab, rust-inhibiting enamel, TT-E-485. Aluminum portions, including the cylinder fins, are left unpainted. The base of the cylinders is coated with waterproofing, electrical, ignition varnish, MIL-V-13811.

c. Engine Compartments. All exterior surfaces of combat vehicle engine compartments not exposed to outside view shall be painted white or light green for better reflection of light.

d. Radiators on Liquid-Cooled Engines. When radiators are stripped to bare metal, an external coat of radiator paint, NSN 8010-00-728-8228, should be applied to outside of tanks and core after they have been assembled. Ensure paint does not restrict air flow through core elements, since clogged core elements will not dissipate heat.

3-17. TAPE TEST (PAINT ADHESION TEST) FOR ALL FINISHES

(See figure 3-1.) Test paint adhesion on coated surfaces. Do not use test panels instead of actual production units. Test the surface after the paint finish has cured on an out of the way place acceptable to quality assurance representatives as follows:

a. Make a V-shaped scratch through the paint finish with a sharp metal blade. Make the scratch approximately two (2) inches long and one half (1/2) inch between edges at the widest point.

b. Press a piece of pressure sensitive tape (3M code no. 250 or equivalent) firmly over the V, in the direction indicated on figure 3-1. Press out air pockets.

c. Wait at least ten seconds, then quickly pull the tape away, in the direction indicated in figure 3-1.

d. Interpret test results as follows:

(1) If no paint comes off of the taped area, the coating is acceptable. (Removal of overspray (para 4-13a) is not a test failure.)

(2) If the topcoat, primer or pretreatment comes off with the tape, then the coating has failed the test and must be removed and another coating applied and tested.

(3) After test is passed, repair the scratched area by feathering-in with appropriate pretreatment, primer and topcoat.

2"

½"

SCRIBE/CUT INTO PAINT

PAINT SAMPLE/AREA TEST

TAPE

APPLY TAPE
IN THIS
DIRECTION

PULL TAPE IN
THIS DIRECTION

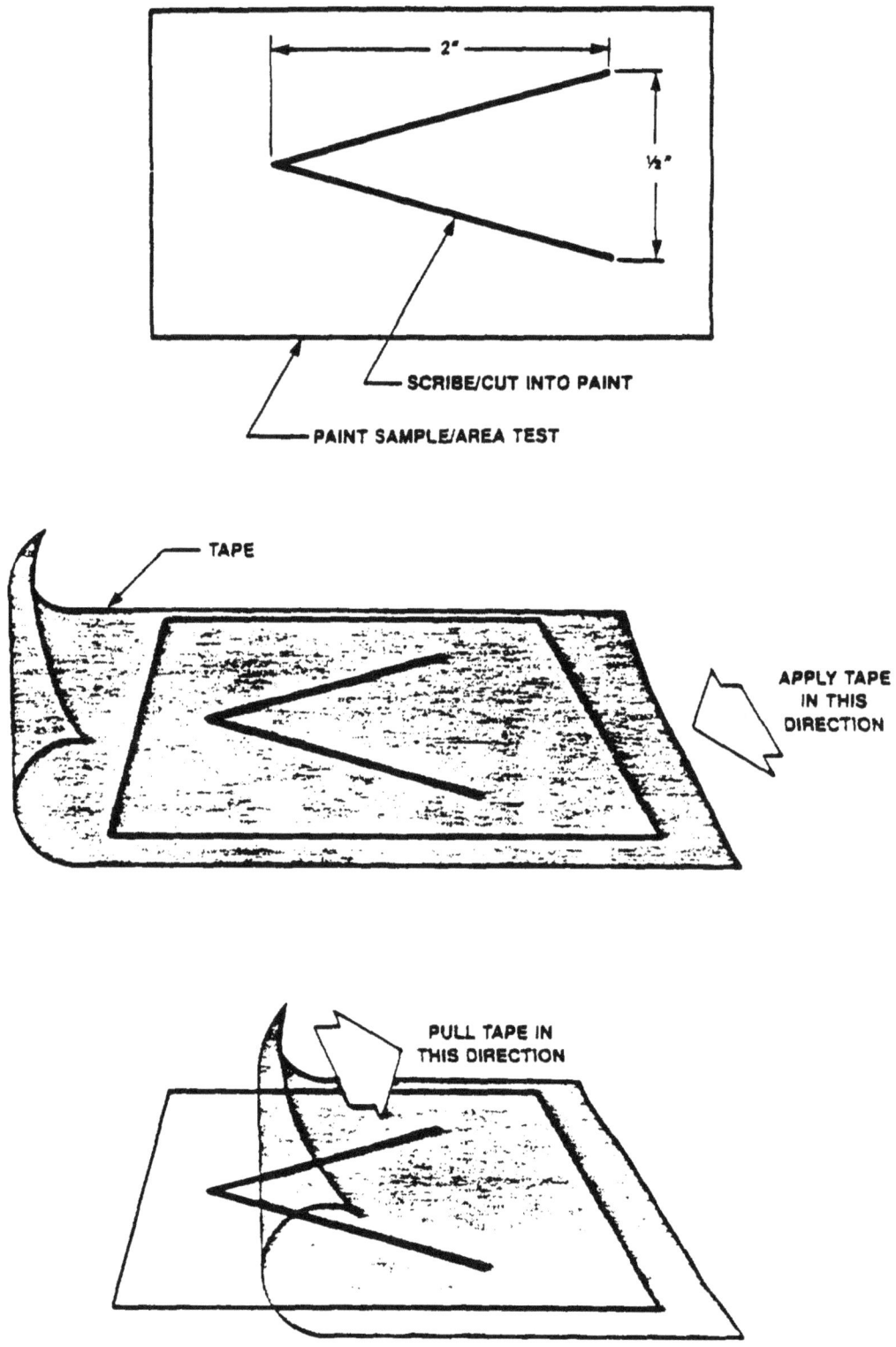

Figure 3-1. Tape Test.

CHAPTER 4
CAMOUFLAGING PROCEDURES

Section I. GENERAL

WARNING

Before beginning any painting-related activity, read Chapter 1, Section II, Safety Summary.

4-1. CAMOUFLAGE PATTERN PAINTING.

a. This chapter covers methods of applying Camouflage Paint Patterns (CPP) to Army materiel. It also covers procedures for inspecting applied CPP. Equipment consists mainly of brushes, rollers and spray guns; finish systems consist of Chemical Agent Resistant Coatings (CARC), such as those listed in table 3-3. The CARC topcoats are suited for camouflage painting and protect military materiel against visual and infrared detection and chemical agent contamination. The CARC camouflage topcoats are MIL-C-46168 and MIL-C-53039. The CARC epoxy primers are MIL-P-23377, MIL-P-53022, and MIL-P-53030. Appendix B contains tables with NSNs for these materials.

b. Information on specific finish systems and kinds of paint to be used for various applications is contained in Chapter 3. Surface preparation is also covered in Chapter 3. Application techniques are covered in Chapter 5. For additional information on camouflage, refer to the following publications:

AR 750-1   Maintenance of Supplies and Equipment, Army Materiel Maintenance Policies
FM 5-20   Camouflage
TM 5-200   Camouflage Materials

c. Major items to be camouflage painted are weapons systems, vehicles, communications equipment, construction equipment, and materials handling equipment. The painting procedures described in this chapter do not apply to aircraft. Three-color camouflage patterns are created for all combat, combat support, and combat service support equipment having an area greater than 9 square feet on one or more sides. New items of tactical equipment normally will be CARC coated in a three-color camouflage pattern at the factory or depot. Some items, however, may be painted with lusterless CARC green 383 when the pattern has not been developed. These items may be three-color camouflage patterned, at the commanding officer's discretion, any time after CPP design development.

d. Only Intermediate and Depot level personnel with equipment and paint booths meeting OSHA standards are authorized complete painting and/or repainting of equipment or components; if such equipment/booths are not available, only touch-up efforts are authorized. Unit personnel are permitted to use topcoats and primers for touch-up efforts only,

4-2. REASON FOR PATTERN PAINTING

a. All military vehicles and equipment have characteristic shapes and shadows. These shapes and shadows contrast with the material surroundings and make the object stand out. Pattern painting using wavy, irregular patches of camouflage colors does much to break up the characteristic shapes of the equipment by reducing contrasts with soil and vegetation, pattern shape, and placement. Patterns have been designed for each type of vehicle to cut off sharp corners, avoid straight, vertical, and horizontal lines, and extend shadows in shapes similar to natural features and vegetation; however, the accuracy with which the CPP is applied completely determines how well the CPP camouflages the equipment.

b. Pattern painting is not a magic, cure-all camouflage technique, but it makes the item much harder to see and recognize as a military object. It also provides an excellent base for applying further natural camouflage such as tree limbs, shrubs, and grass.

NOTE

Mixing CARC colors with one another will alter their individual effectiveness when applied to the end item. Blending, therefore, is not permitted.

c. Camouflage finishes better lend themselves to touchup painting than do the current olive-drab coatings. Slight mismatches in color are expected at times and will not be noticeable except upon close inspection. Likewise, minor abrasions and scaling of surfaces will be equally inconspicuous. Marring and surface lightening due to handling is characteristic of camouflage coatings and does not impede camouflage or infrared properties. This is typical of low gloss and sheen coatings in dark colors and is considered satisfactory.

Section II. PATTERN APPUCATION

WARNING
The local safety office and industrial hygienist must be consulted before beginning/changing any painting operation.

4-3. PATTERN DESIGNS

Before applying camouflage paint patterns (CPP) to equipment, the pattern design for that equipment must be obtained.

Each type of item has its own CPP design which consists of 5 views of the equipment: front, back left side, right side and top. Also issued with the designs are inspection worksheets and overspray gauges, which are necessary for inspecting CPP once it is applied. To obtain design, inspection worksheets, and/or overspray gauges, write to:

Commander
Belvoir Research, Development and Engineering Center
ATTN: STRBE-JDA
R. Belvoir, VA 22060-5606

For the equipment being camouflaged, furnish the following data with your request:

- National Stock Number (NSN)
- Nomenclature
- Model Number (where applicable)
- Standard Study Number (if known)
- Line Item Number (if known)

See figure 4-1 for an example of the CPP for the M 113 Personnel Carrier. Within each area is a number that stands for the color to be painted. The base, or primary" color is designated #2. Color #2 is usually applied first over all surfaces. Color patches #1 and #3 are then applied over the primary coating.

ARMORED
PERSONNEL CARRIER
FULL TRACKED M 113

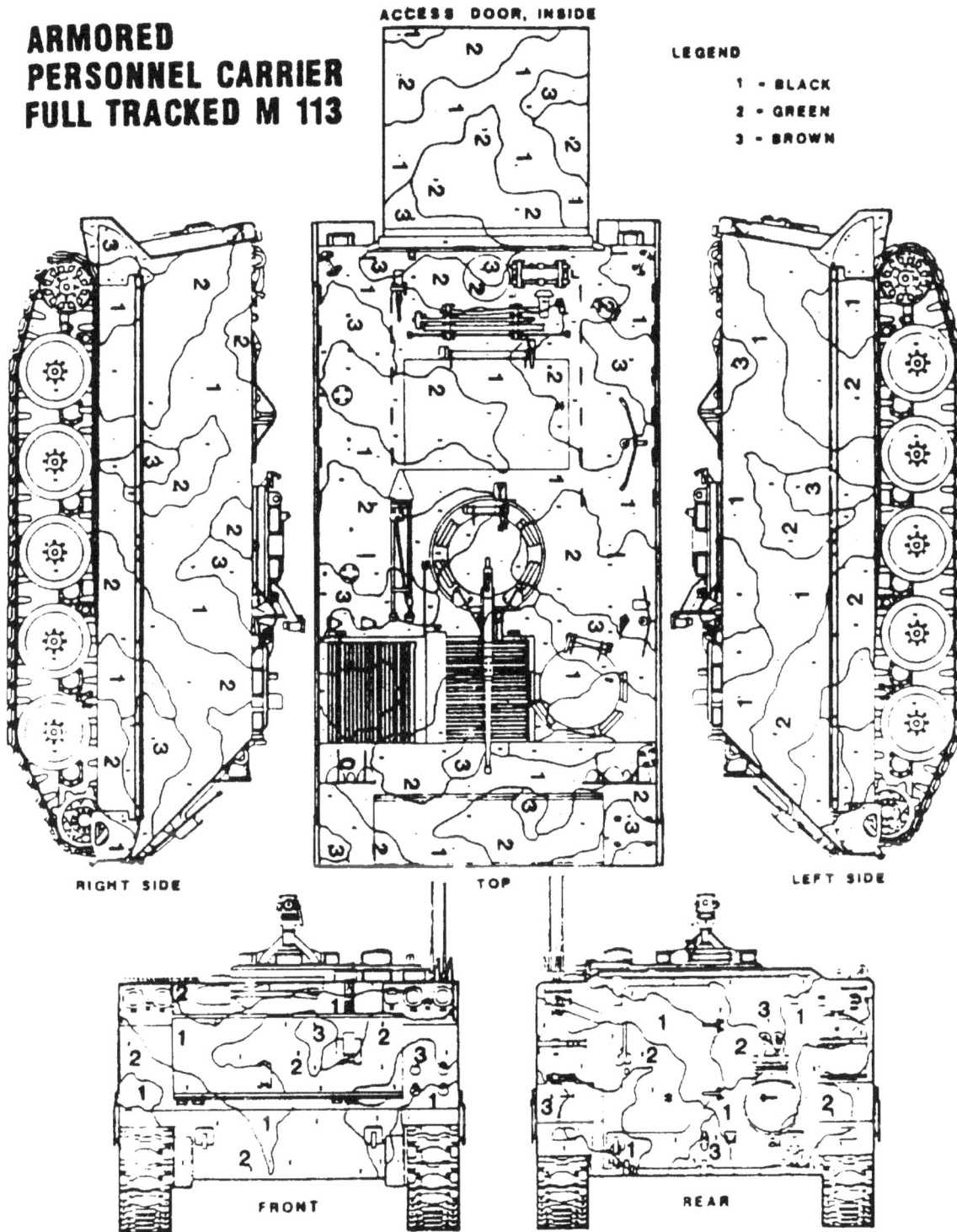

ACCESS DOOR, INSIDE

LEGEND

1 - BLACK
2 - GREEN
3 - BROWN

RIGHT SIDE

TOP

LEFT SIDE

FRONT

REAR

Figure 4-1. Pattern Painting Design for the M113 Personnel Carrier

4-4. CHOICE OF METHOD

There are four CPP application methods: robotic, template, projection and manual. Robotic application is the most accurate and consistent, and where a number of like items must be camouflaged, this method merits serious consideration. Template application is the next most accurate and repeatable method, and where the robotic method is deemed impractical, the template method should then be considered. The projection method is less accurate and repeatable; however, where a limited number of like items are to be camouflaged, it may be more practical/cost-effective than the robotic and template methods. Manual application is the least accurate, least repeatable method, and for this reason its use is greatly discouraged. It should be used only when the other three methods, for whatever reason, have been ruled out.

4-5. ROBOTIC METHOD

The robotic method of applying patterns uses an automated robotic program to establish the color boundary lines simultaneously as the paint is applied to the bands and patches. Each color is automatically applied to its respective color area by robotics preprogrammed to apply the designated pattern. No human participation is required.

   a. *Degree of Accuracy/Consistency.* It is considered to be the most accurate and repeatable method of CPP application.

   b. *Inspection Requirements.* The first pattern applied by a robotic program should be fully inspected (see Section III, CPP Inspection Procedures); however, once that pattern passes inspection, that robotic painting program is 'certified'. Subsequent patterns applied using certified programs need be inspected only on a random sample basis.

4-6. TEMPLATE METHOD

The template method of pattern application uses either rigid or soft templates to locate and mark the pattern color boundary lines on an item that requires a CPP. Templates are fabricated from rigid material (wood, aluminum, etc.) or soft, flexible material (mylar, plastic sheets, etc.). The fabricated template is precisely positioned on the surface to be patterned, which must already be completely coated with the base color (#2 on the CPP design). The boundaries are then located and drawn with soapstone or chalk onto the surface. Finally, the painter fills in color areas #1 and #3 of the CPP with the colors designated on the design.

   a. Degree of Accuracy/Consistency. After the robotic method, it is the next most accurate and repeatable.

   b. Inspection Requirements. The first pattern applied using a template must be fully inspected in accordance with Section III, Inspection Procedures; however, once that pattern passes inspection, the template is 'certified'. Subsequent patterns applied with this template need be inspected only on a random-sample basis.

4-7. PROJECTION METHOD

The projection technique utilizes transparent reductions of the CPP drawings which are transferred directly onto the item by illuminated projection. This technique permits the color boundary lines to be traced manually using chalk or soapstone.

   a. *Degree of Accuracy/Consistency.* It is an inaccurate, inconsistent method of CPP application, but it is slightly more desirable than the manual pattern application technique. Its use, however, may be necessitated by practicality and cost effectiveness constraints, especially when there are only a small or limited number of the same item requiring the CPP application (i.e., when the cost tradeoff may not justify the development of soft or hard templates or robotic programs).

b. Inspection Requirements. Every pattern applied using the projection method must be fully inspected in accordance with Section III, Inspection Procedures. No random sampling is permitted.

4-8. MANUAL METHOD

The manual application of color boundary lines is considered to be the least accurate and least repeatable method of CPP application. This method uses a modified "free-hand" approach for applying the color boundaries with the aid of various pattern point guideline techniques such as grid layout, transfer of scaled points from CPP drawings, etc.

a. Degree of Accuracy/Consistency. The manual method is the least accurate, least repeatable application method. Therefore, its use is greatly discouraged. It is considered a last resort' method.

b. Inspection Requirements. Every pattern applied using the manual method must be fully inspected in accordance with Section III, Inspection Procedures. No random sampling is permitted.

4-9. MARKINGS ON CAMOUFLAGED EQUIPMENT

After the equipment has been pattern painted, only the following markings are to be applied:

NOTE

Markings on CARC camouflaged equipment will be CARC in colors designated in figure 4-2.

a. Unit Identification. Type and location remain the same.

b. National Symbol. Paint a 3-inch star on both front and rear. Symbol should be centered on the equipment, on line with unit ID markings. On the rear of wheeled vehicles, the star may be placed on the tailgate.

c. Agency Identification and Registration Number. The identification and registration number shall be placed on any appropriate interior area, if available, which is visible from outside a locked or secured item.

*d. **Safety and Instructional Markings.*** Markings such as tire pressure, fuel type, and fill level will be in letters no larger than one inch. Markings directly related to troop safety, such as wrecker boom capacity and danger zones, must be evaluated by safety personnel.

e. CARC Markings. Equipment with data plates shall have the word "CARC" painted in a conspicuous area as near the data plate as possible. Painting shall be in block letters, as large as possible, not to exceed one inch. All major items having a log book shall have the Equipment Control Record, DA Form 2408-9, annotated in the lower left corner of Block 21, Remarks, reflecting the word "CARC", and the date applied.

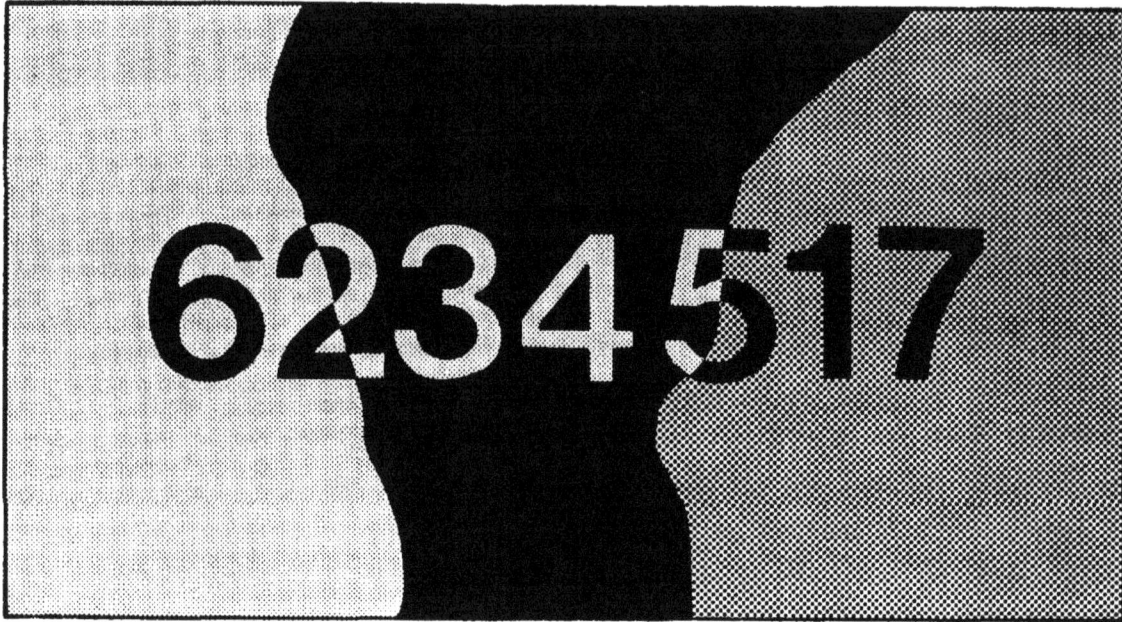

BACKGROUND COLOR AREA	LUSTERLESS LETTER COLOR
BROWN 383	BLACK 37030 or 37038
GREEN 383	BLACK 37030 or 37038
BLACK 37030 or 37038	GREEN 383
WHITE	BROWN 383
TAN 686	BROWN 383

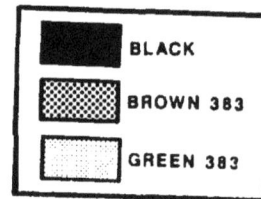

BLACK
BROWN 383
GREEN 383

Figure 4-2. Markings on Camouflaged Equipment.

Section III. INSPECTION PROCEDURES

4-10. GENERAL

The full CPP inspection consists of three levels of inspection: Level I, Level II, and Level III. Level I is essentially a visual conformance check from 50 feet or more. Level II is a close-up boundary inspection using a CPP overspray gauge. Level III is a detailed comparison of actual physical measurements to nominal dimensions.

4-11. MATERIALS/TOOLS REQUIRED

After becoming completely familiar with the inspection procedures outlined below, the inspector should assemble the following materials/tools:

- CPP designs, including inspection worksheet (para 4-3) and overspray gauge (fig 4-3)
- One (1) 6-foot steel measuring tape, graduated in sixteenths (or thirty-seconds) of an inch
- One (1) 6-foot steel measuring tape, graduated in tenths of an inch
- Small T-square and/or straight edge
- Soapstone or chalk
- Pen, pencils, colored pencils
- Paper/notebook
- One plumb bob, with string

- CPP designs, including inspection worksheet (para 4-3)
- One (1) 6-foot steel measuring tape, graduated in sixteenths (or thirty-seconds) of an inch
- One (1) 6-foot steel measuring tape, graduated in tenths of an inch
- Small T-square and/or straight edge
- Soapstone or chalk
- Pen, pencils, colored pencils
- Paper/notebook
- One plumb bob, with string
- Overspray gauge (figure 4-2)

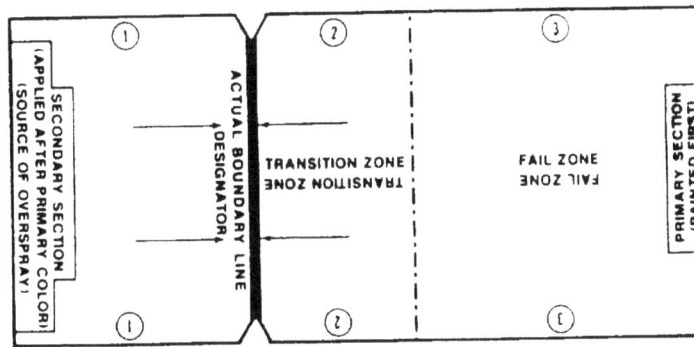

Figure 4-3. Overspray Gauge

4-12. LEVEL I INSPECTION

In this level of inspection, the inspector, from a distance of 50 feet or more from the item, visually compares the colors, shapes and boundaries of the applied pattern to those of the design pattern, by using the following sequence of steps:

a. Ensure the data on the identification plate of the item to be inspected corresponds to the data on the CPP drawings. If not, the CPP inspection cannot be performed until the correct drawing set is obtained (para 4-3).

b. With 2 different colored pencils, shade color areas #1 and #3 of the drawing set, including inspection worksheet.

c. Begin with the right side view. If from that view the items 20 feet or less long, stand 50 feet from it. If item is over 20 feet but less than 40 feet in length, stand 75 feet from it. If item is longer than 40 feet. stand 100 feet from it.

d. Compare the right side as drawn with the actual right side view of the item:

(1) Ensure all black disruptive bands have been properly located on the item.

(2) Verify comparative size and direction of black bands with those on drawing.

(3) Ensure black boundary lines closely conform to the shape of boundaries depicted on CPP drawings.

(4) Ensure all brown patches have been properly located on item.

(5) Ensure contours are sharply defined, and shapes of brown patches closely conform with those on CPP drawings.

e. Document any failures (differences found between applied pattern and drawing pattern) as follows:

(1) Mark failed area on item with chalk or soapstone.

4-7

(2) Record failure on inspection worksheet.

(3) Describe failure in notebook for painter/contractor debriefing.

(4) Report failure in specific detail to painter/contractor to facilitate correction. Reference points may be used to show failure location.

(5) Provide constructive guidance on corrective action (e.g., inspector may even draw correct contour line on item and direct repainting within this line).

f. Repeat c through e for the other four drawing views. Level I inspection is complete when each view is examined in this manner, and when all corrections have been witnessed and accepted by inspector.

4-13. LEVEL II INSPECTION

In this level of inspection, the borders are examined to ensure color definition. The contrast between color areas is key to CPP effectiveness, and, for this reason, overspray of one color into another color area must be minimized to a 1-1/2 inch tolerance. An overspray gauge (figure 4-3) is used to determine whether overspray on the applied CPP is within this narrow tolerance.

a. Overspray. Unless applied robotically, pattern colors are applied in sequence, beginning with a base coat in the primary color (#2 on CPP drawings). The other two color patches/bands are spray-painted, one after the other, over this base coat. A normal result of this process is overspray, the overlapping mist of one paint color on an adjacent color area which impairs CPP effectiveness. With experience, painters can minimize overspray.

b. Overspray Gauge. The overspray gauge (figure 4-3) is a card-like, transparent sheet of plastic with white markings (to contrast with camouflage colors) which indicate the tolerance, or 'transition' zone. The gauge is overlayed on the border area; the actual border line is then located and aligned under the gauge boundary line DESIGNATOR, and the overspray is examined. When overspray extends into the 'FAIL ZONE', the failure location is noted by marking the notches at both ends of the gauge boundary line DESIGNATOR with chalk or soapstone.

c. Areas to be Inspected. Level I inspection included a visual check from a distance of contour definition. Any areas noted where contours were NOT sharply defined should be inspected for overspray failure. Special attention should be given to black disruptive bands, which are especially important to CPP effectiveness.

d. Level II Inspection Procedure. For each area noted as in sub para c, complete the following steps:

(1) Lay overspray gauge over border, so that FAIL ZONE lays over color area applied first.

(2) Align DESIGNATOR, or boundary line marking, as close as possible to fog/boundary transition area, so that it is just short of overlapping spray, but so that no primary color can be seen in secondary section.

(3) Examine overspray. If it extends into FAIL ZONE, record Level II failure as follows:

(a) Mark notches at both ends of DESIGNATOR line on gauge with chalk or soapstone.

(b) Record overspray failure on inspection worksheet.

(C) Make descriptive entry in notebook for painter/contractor debriefing.

(d) Report overspray failure in specific detail to painter/contractor. Show failure location.

(e) Provide constructive guidance on corrective action.

4-8

NOTE

Dimensions are not labeled "LD" and "BW" as shown here on actual inspection worksheets.

NOTE

RP and black border not in same horizontal plane.

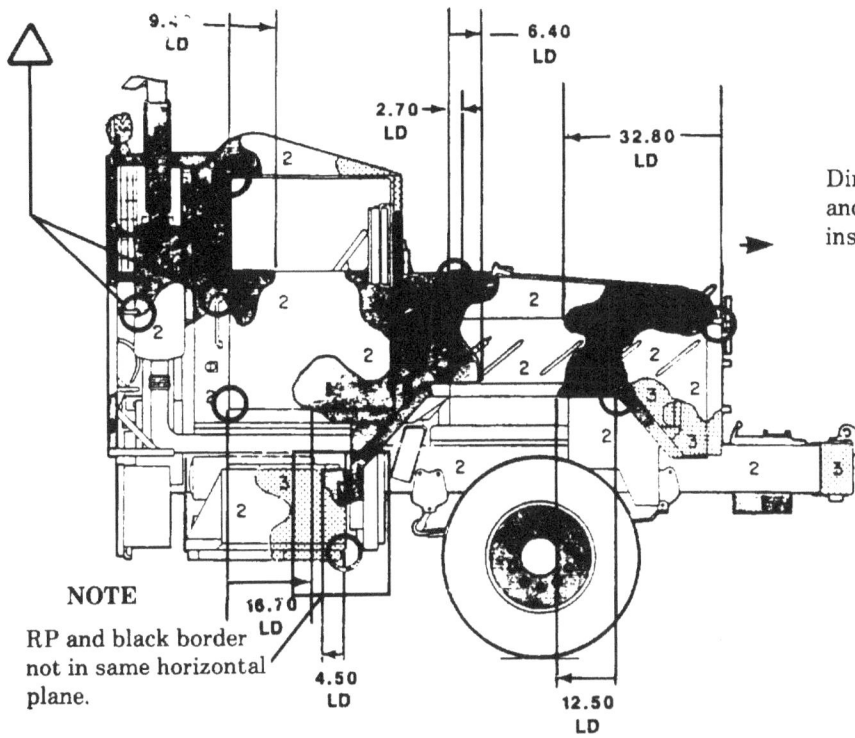

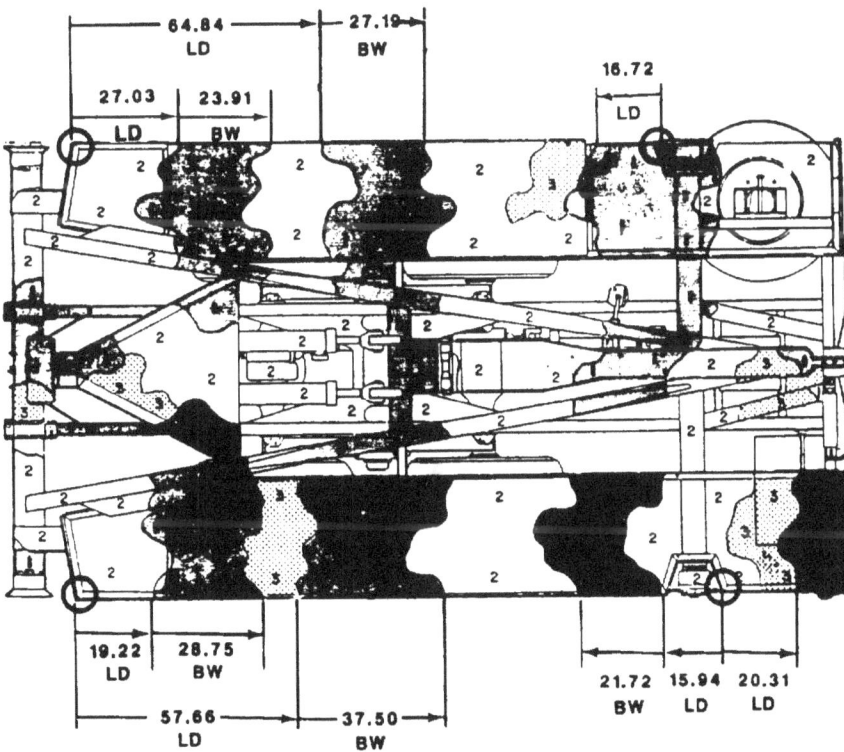

Figure 4-4. Typical Views on Inspection Worksheet.

e. Level II Inspection Completion. This level of CPP inspection is complete when each view of the item is inspected in this manner, and when inspector has witnessed and accepted all corrections.

4-14. LEVEL III INSPECTION

This level of inspection entails contrasting actual versus nominal dimensions of the CPP. Specifically, the inspector verifies that all black disruptive bands were applied within + 1 inch of the location specified on the worksheet. Reference points are used to make these measurements. At least one point on each boundary of the black bands must be measured from a nearby reference point. Inspections, however, are not limited to one point per border, nor to the reference points indicated on the worksheet.

a. Reference Points (RPs). RPs are denoted on inspection worksheets (figure 4-4) by small circles (o). They are usually (not always) located at the intersection of two physical line features, such as the corner of a door.

b. Deltas. Deltas are denoted by triangles (A) on the inspection worksheets (figure 4-4). They are reference points which are located within an inch radius of a black band border.

c. Locating Dimensions (LDs). The LD is the horizontal distance between a reference point and a black band border. The LD must be within + 1 inch of the dimension specified on the inspection worksheet (figure 4-4).

d. Bandwidths (BWs). The BW is the distance between a point on one border of a black band and a point on a different border of the same band. Like the LDs, actual BWs must be within + 1 inch of those specified on the inspection worksheet.

e. Level III Inspection Procedure. The simplest way to inspect the CPP at this level is to check each delta, LD and BW indicated on the worksheet.

(1) Deltas. A delta on the worksheet (figure 4-4) is indicated by a small triangle. Check these as follows:

(a) Locate delta on worksheet. Some physical equipment feature, such as the top of a door handle or a panel corner, should intersect a black band border within ± 1 inch.

(b) Go to equipment being inspected, and locate this feature.

(c) On the applied CPP, locate the corresponding black band border, and using the overspray gauge, find the actual border line, marking the notches with chalk or soapstone (para 4-13d(2)).

(d) Use a ruler to determine whether any point on the actual border line is within a 1 inch radius of the actual physical feature.

(e) If no point on the border line is within this radius, a level III failure must be documented. As with level II failures, the failure location on the applied CPP should be marked with chalk or soapstone, and the failure should also be recorded on the worksheet. A notebook entry and detailed report to painter, as with level II failures, should also be made. The inspector must guide, witness and accept corrections of failures. Follow documentation procedure in para 4-13d(3).

(2) LDs. A LD on the inspection worksheet (figure 4-4) is a dimension (in inches) with arrows and lines from a RP, denoted by a small circle, to a point on a black band border. Check LDs as follows:

(a) Locate LD on worksheet. Note the RP and the point on the black band border.

(b) Go to equipment, and locate the equipment feature which corresponds to the RP on the worksheet.

(c) On the applied CPP, locate the corresponding black band border.

(d) If the border is at some point straight across from (on the same horizontal plane as) the RP, use the overspray gauge to locate the actual border line (para 4-13d) where border intersects RP's horizontal plane.

(e) Mark point on border line where it intersects this plane, and measure distance from this point to RP.

(f) If border is not on same horizontal plane as RP (see figure 4-4 for example), drop a plumb bob from the higher point, and measure distance from lower point to that point where its horizontal plane intersects plumb line. (Again, the overspray gauge can help distinguish the actual border line (para 4-13d(2).)

NOTE

Vehicle must be level to get accurate measurement.

(g) This distance must be within + 1 inch of the LD on worksheet. If it is not, document a level III failure in same way as level II failures (para 4-13d(3)(a)-(g)).

(3) BWs. A BW on the inspection worksheet (figure 4-4) is a dimension (in inches) with arrows and lines from a point on a black band border to a point on an opposite border of same black band. Check these as follows:

(a) Locate BW on worksheet. Note points on opposite borders of same black band.

(b) Go to equipment and locate corresponding points on actual applied borders.

(c) Use overspray gauge to distinguish each actual border line (para 4-13d(2)).

(d) Measure distance between these lines at points corresponding to those marked on worksheet.

(e) If this distance is not within + 1 inch, document level III failure in same way as level II failures (para 4-13d(3)(a)-(g)).

f. Level III Inspection Completion. This level of CPP inspection is complete when at least one point on each black band border in each view of the equipment has been checked to ensure its location is within ± 1 inch of that specified on the drawings. Checking each LD, delta and BW on the inspection worksheet minimally fulfills this requirement; however, additional measurements may be made to increase confidence levels at discretion of inspector. Inspector must witness and accept correction of all failures found.

4-15. COMPLETION OF FULL CPP INSPECTION

The CPP inspection is complete when all three levels of inspection are complete.

4-16. SUMMARY OF CPP INSPECTION REQUIREMENTS

Inspection requirements vary with the method of CPP application. As stated in Section II, robotic and template methods require the first applied pattern to pass inspection (all levels), thereby making that robotic program or template certified. Subsequent patterns applied using that program or template shall be inspected on a random sample basis. Manual and projection methods require each and every applied pattern to be inspected (all levels).

4-11/(4-12 blank)

CHAPTER 5
APPLICATION TECHNIQUES AND EQUIPMENT

Section I. SPRAY GUNS AND PRESSURE CANS

WARNING

Before beginning any painting-related activity, read Chapter 1, Section II, Safety Summary.

5-1. SPRAY GUN APPLICATION

a. Use. Spray gun equipment can be used for any type of finish and on any surface. It does not replace the brush for certain operations, yet there are definite types of work it can do more easily and better than the brush. The spray gun is obviously a tremendous time-saver and its use is recommended when a large volume of work is encountered. The spray gun is particularly adaptable to touchup and maintenance work when the ability to blend old and new surfaces is important. Spray application of any finish type requires respiratory equipment.

b. Training. The proper operation of spray guns and auxiliary equipment is not difficult to learn, but the necessity exists for training operators. Only through such training can the full flexibility and operation of spray guns be realized.

5-2. SELECTION OF SPRAY GUNS

a. Definition. A paint spray gun (see figure 5-1) is a mechanical means of bringing compressed air and paint together, atomizing or breaking up the paint stream into a spray, and ejecting it for the purpose of applying a coating.

b. Types. There are two types of spray guns: attached containers and separate containers. These types can be further subdivided into bleeder and non bleeder, external and internal mix, and suction and pressure feed types. The commercially available pressure cans belong to the attached container type. They have a limited use for touchup where compressed air is inaccessible to the job. Airless, portable spray equipment that requires an electric connection also belongs to this type.

(1) *Bleeder and nonbleeder guns.*

(a) A bleeder-type gun is characterized by an intentional continuous leakage of air from some part of the gun. This prevents building up air pressure within the hose and permits its use with small compressing systems that are not equipped with an automatic pressure-controlling device. The trigger in a bleeder-type gun controls only the flow of fluid.

(b) A nonbleeder-type gun is one in which the trigger controls the passage of both air and fluid. Some type of pressure-controlling device must be used with it.

(2) *External and Internal Mix.*

(a) An external-mix gun is one which mixes air and fluid outside the air cap.

(b) An internal-mix gun mixes air and fluid within the air cap.

NOTE

The term internal mix applies to the air cap alone.

(3) *Suction and Pressure Feed.*

(a) A suction-feed gun is designed to feed the fluid into the air stream through a vacuum created by raising the fluid tip above the air cap. Generally, guns of this type are used with quart-size or smaller containers.

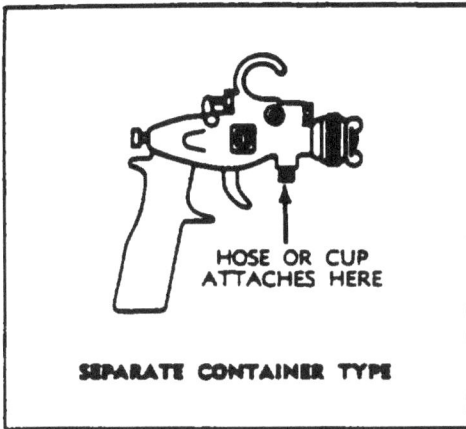

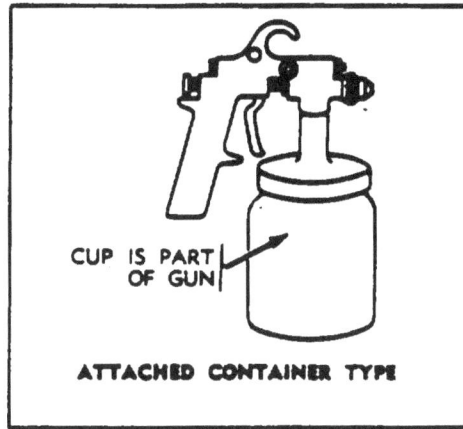

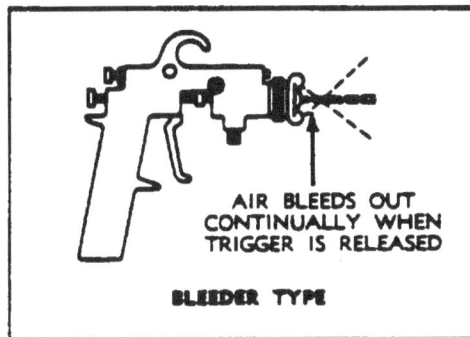

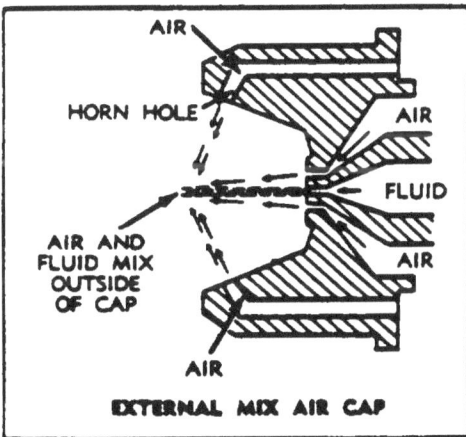

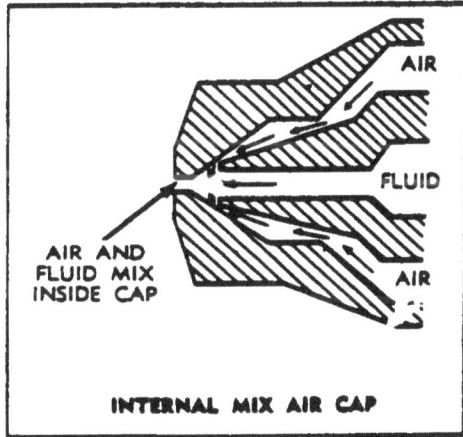

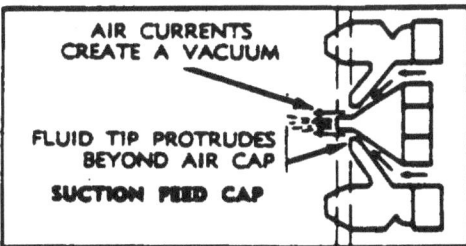

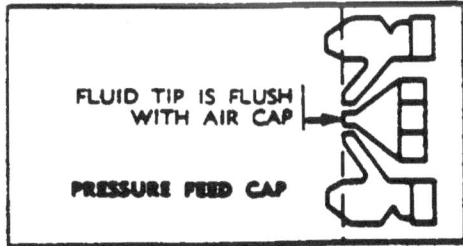

Figure 5-1. Spray Gun Types

(b) A pressure-type gun feeds fluid into the air stream (the air cap and fluid tip are flush), by means of applied air pressure that forces fluid from the container to the gun.

(4) Airless Spray Equipment.

Airless spray equipment uses an electrically operated vibration element which forces the paint up through a tube from the bottom of the container to a nozzle in the cover. This type can be recharged with paint See the manufacturer's instructions for the type of current and voltage required.

5-3. SELECTION OF AIR CAPS. NEEDLES. AND NOZZLES

The performance of an air gun with any kind of material depends primarily on the selection of the proper air cap, fluid needle, and fluid tip (or nozzle). Manufacturers identify combinations of these parts intended to be used together, and their recommendations should be followed in respect to the proper combination for a particular material. Occasionally, changing the type of feed will necessitate a different combination of air cap, fluid tip, and fluid needle.

5-4. SPRAY GUN TECHNIQUES

WARNING

All personnel who work in a spray painting booth must wear a NIOSH approved respirator as well as personal protective equipment (PPE), when spray painting operations are underway. Refer to para 1-7.

WARNING

Spray painting will be done only in areas designated for that use. All personnel within a paint booth must be equally protected with proper PPE in accordance with para 1-7. No unprotected personnel may enter a paint booth without protection until 30 minutes after all painting/cleaning is completed.

WARNING

Only one person will spray paint on an item at a time, unless all people are protected in accordance with para 1-7. This is to eliminate the hazard of accidently spraying paint on another person.

Figure 5-2. Deleted

ARCING CAUSES
UNEVEN APPLICATION

6 TO 10 INCHES

-2-
MOVE GUN IN
STRAIGHT LINE

DO NOT
ARC STROKE

-1-
BEGIN STROKE
THEN PULL TRIGGER

-3-
RELEASE TRIGGER
BEFORE COMPLETING
STROKE

CORRECT METHOD

WRONG METHOD

SHOWING PROPER METHOD OF MAKING SPRAY GUN STROKE

SURFACE

SURFACE

CORRECT METHOD

WRONG METHOD

SPRAY PAINTING CORNERS

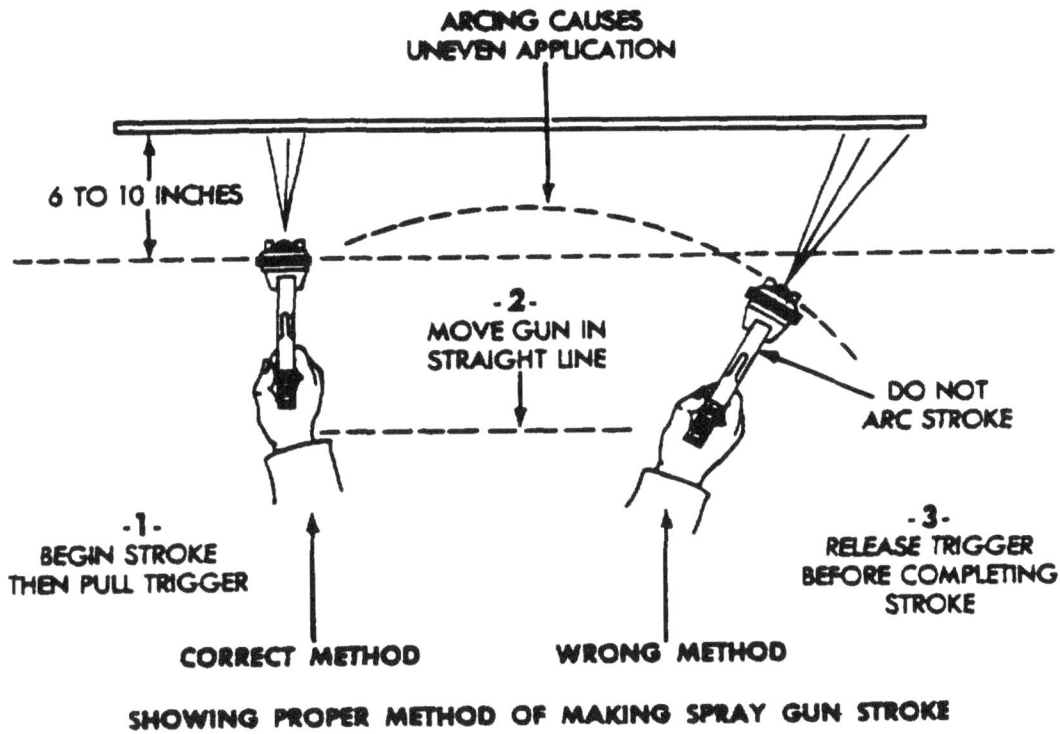

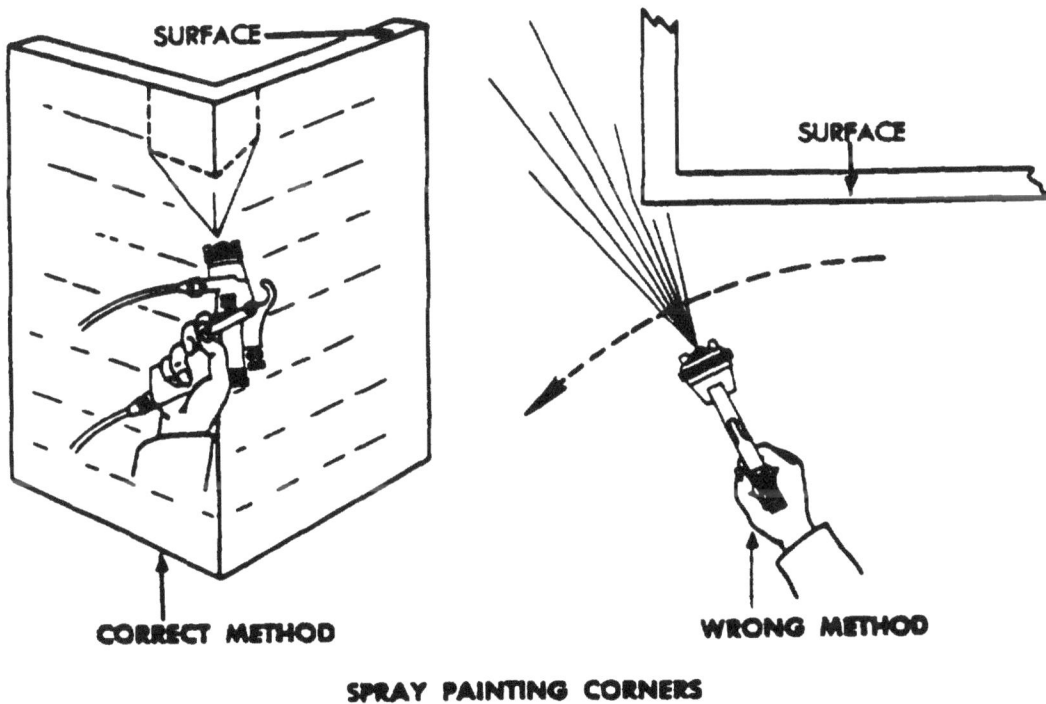

Figure 5-3. Proper Method for Making Spray Glen Strokes

f. Masking. When spraying, cover or mask all parts such as windows, gages, lubrication fittings, instruments, and other parts which are not to be painted.

(1) Small areas or irregular-shaped parts are covered with crepe-backed masking tape. On larger areas, a sheet of paper slightly smaller than the part to be masked is used, with the paper being held in place by a srip of masking tape, which overlaps the edge of the paper, and holds the paper to the surface being masked. Masking tape is available in various widths and is a convenient material with which to work when covering irregular outlines. The method of applying and trimming the tape is shown in figure 5-4.

(2) Liquid or past-like materials are also available to mask out areas where paint is not desired. After the paint is dry, these areas may be wiped or stripped clean.

(3) When spraying vehicle engines, the use of cloth bandages and socks will protect rubber hose, ignition wires, and flexible tubing, and save much time and material. Cut the cloth bandage to fit the object to be covered, allowing for hose clamps and other parts of the engine. Drawstrings at each end, with a string wrapped around the middle of the bandage and tucked under a flap, prevent overspray from striking the protected object. Fit the sock over the ignition wires and distributor cap, and use the drawstring tie to secure it around the base of the distributor.

5-5. LEAKAGE AND CORRECTION

a. Material Leakage from Fluid Needle Packing Nut. This condition is caused by a loose packing nut or by dry fluid needle packing. To remedy, remove and soften the packing with a few drops of light oil. Re-install and tighten packing nut to prevent leakage. See figures 5-5 and 5-6.

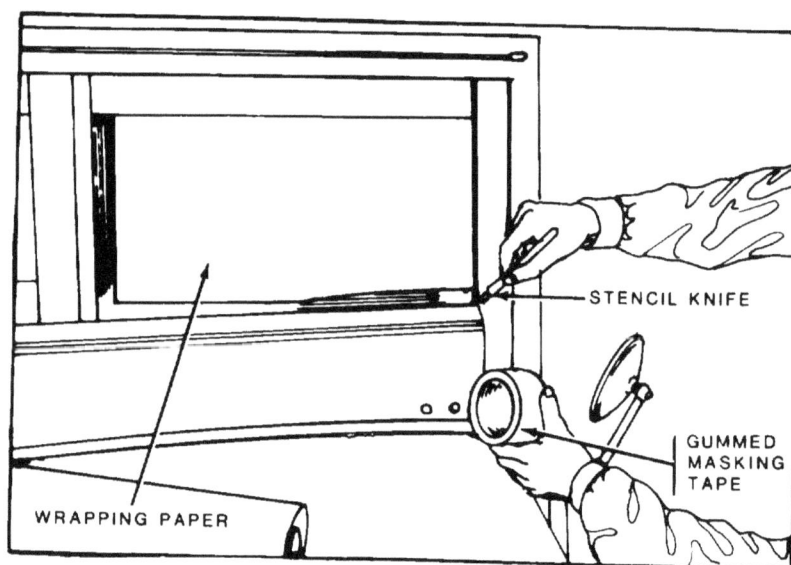

Figure 5-4. Masking

Key to figure 3-5:

1. Air cap retaining ring
2. Air cap
3. Fluid tip
4. Gasket
5. Baffle plate
6. Housing
7. Fluid nipple
8. Packing
9. Packing nut
10. Packing nut
11. Packing
12. Air valve body
13. Air valve
14. Spring
15. Gasket
16. Trigger
17. Air inlet nipple
18. Setscrew
19. Gun body
20. Plunger cylinder
21. Needle plunger
22. Spring
23. Adjusting screw
24. Adjusting nut
25. Adjusting valve
26. Housing
27. Stud
28. Screw
29. Gasket
30. Fluid needle

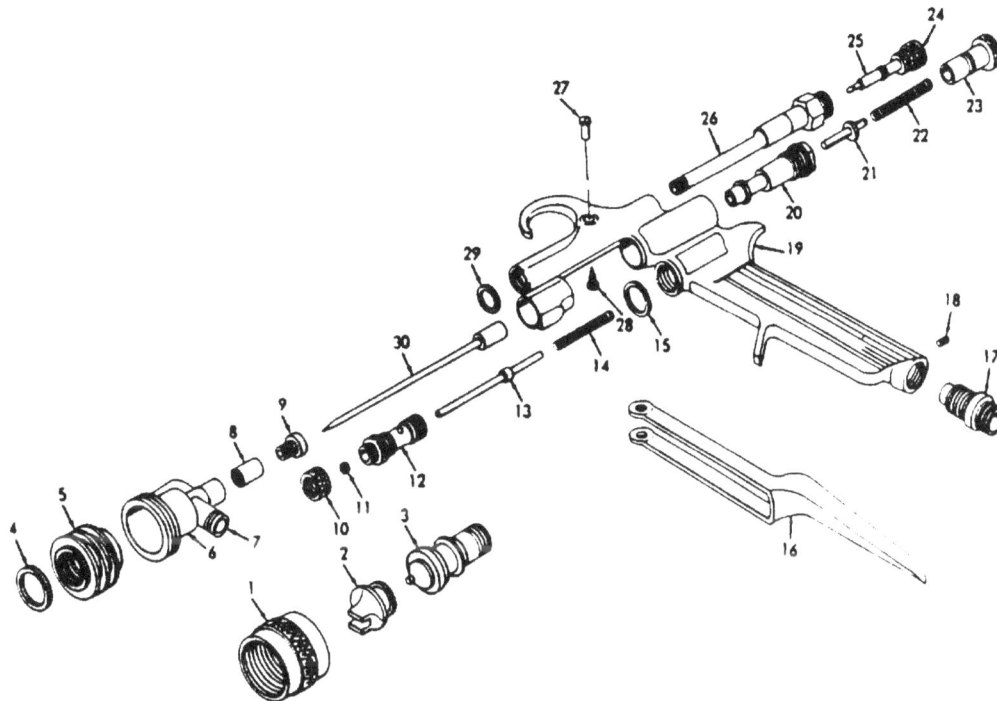

Figure 5-5. Removable Spray-Head Type Spray Gun, Exploded View

5-7

Key to figure 3-6.

1. Air cap retaining ring
2. Air cap
3. Fluid tip
4. Gasket
5. Baffle
6. Gun body
7. Trigger screw
8. Stud
9. Adjustment valve
10. Adjustment assembly
11. Adjustment screw
12. Spring
13. Adjusting screw
14. Fluid needle
15. Air valve plug
16. Air tube
17. Nut
18. Gun handle
19. Air inlet connection
20. Spring
21. Packing nut
22. Packing
23. Fluid nipple
24. Air valve
25. Trigger
26. Air valve body
27. Packing
28. Packing nut

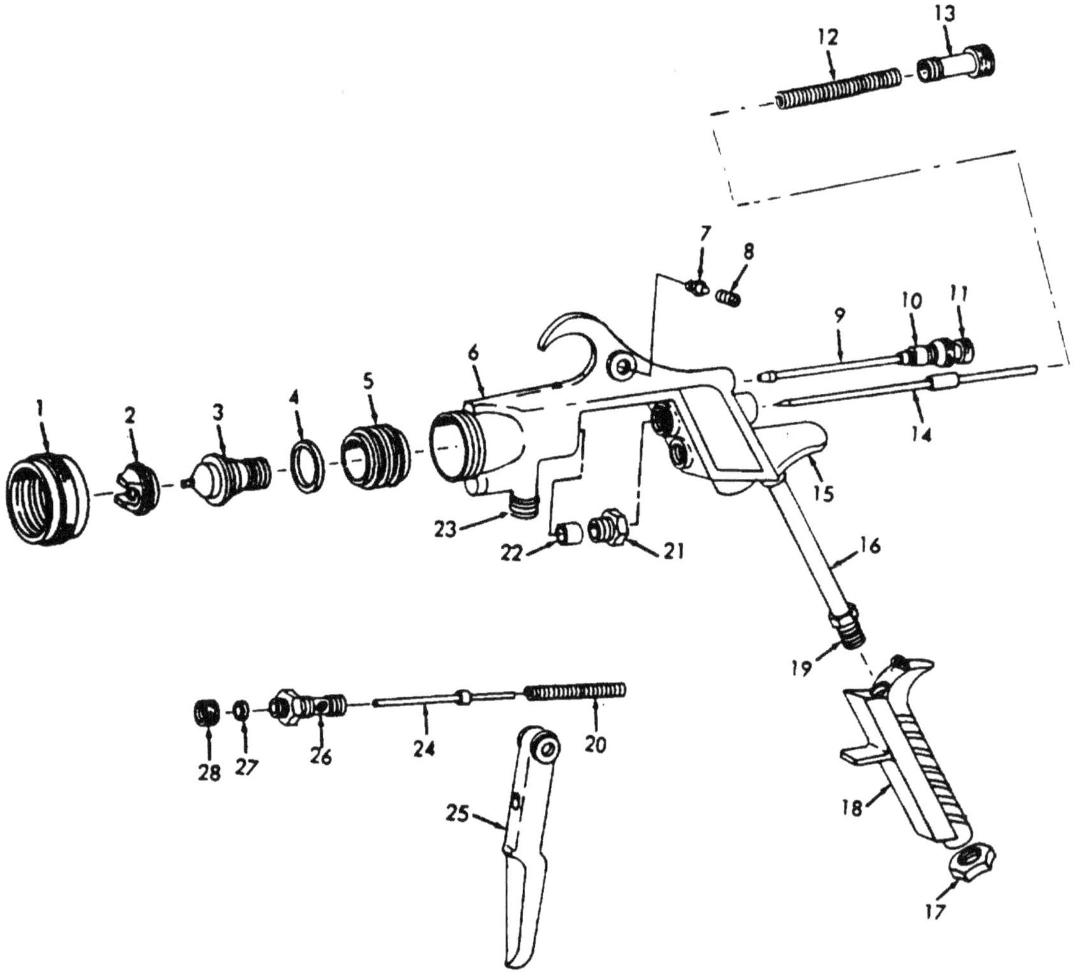

Figure 5-6. Solid-Body Type Spray Gun, Exploded View

b. *Air Leakage from Front of Gun.* This condition is caused by the air valve not seating properly due to:

　(1) Foreign matter on the valve or seat.

　(2) A worn or damaged valve or seat.

　(3) A broken air valve spring.

　(4) A sticking valve stem due to lack of lubrication.

　(5) A bent valve stem.

　(6) A tightly closed packing nut.

c. *Material Leakage from Front of Gun.* This condition is caused by the fluid needle not seating properly due to:

　(1) A worn or damaged fluid tip or needle.

　(2) Lumps of material or foreign matter lodged in the fluid tip.

　(3) A tightly closed packing nut.

　(4) A broken fluid needle spring.

　(5) An improper size needle.

d. *Jerky or Fluttering Spray.*

　(1) In pressure or suction-feed guns, this condition is caused by air leakage into the material line due to:

　　(a) A lack of sufficient material in the container.

　　(b) Tipping the container at an acute angle.

　　(c) An obstructed fluid passageway.

　　(d) A loose or cracked fluid tip in the cap.

　　(e) A loose fluid tip or damaged valve seat.

　(2) Conditions which apply only to suction feed are:

　　(a) Material being too heavy for the suction feed.

　　(b) A clogged air vent in the container lid.

　　(c) A loose, dirty, or damaged fluid inlet connection.

　　(d) The material feed tube (see figure 5-7) is resting on the bottom of the container.

TRIGGER

FLUID NEEDLE
ADJUSTING SCREW

AIR NIPPLE

AIR VALVE
ADJUSTING VALVE
SCREW

FLUID NEEDLE

GUN
BODY

FLUID
NIPPLE

FLUID INLET
CONNECTION

CLOSURE

GASKET

MATERIAL
FEED TUBE

GLASS
CONTAINER

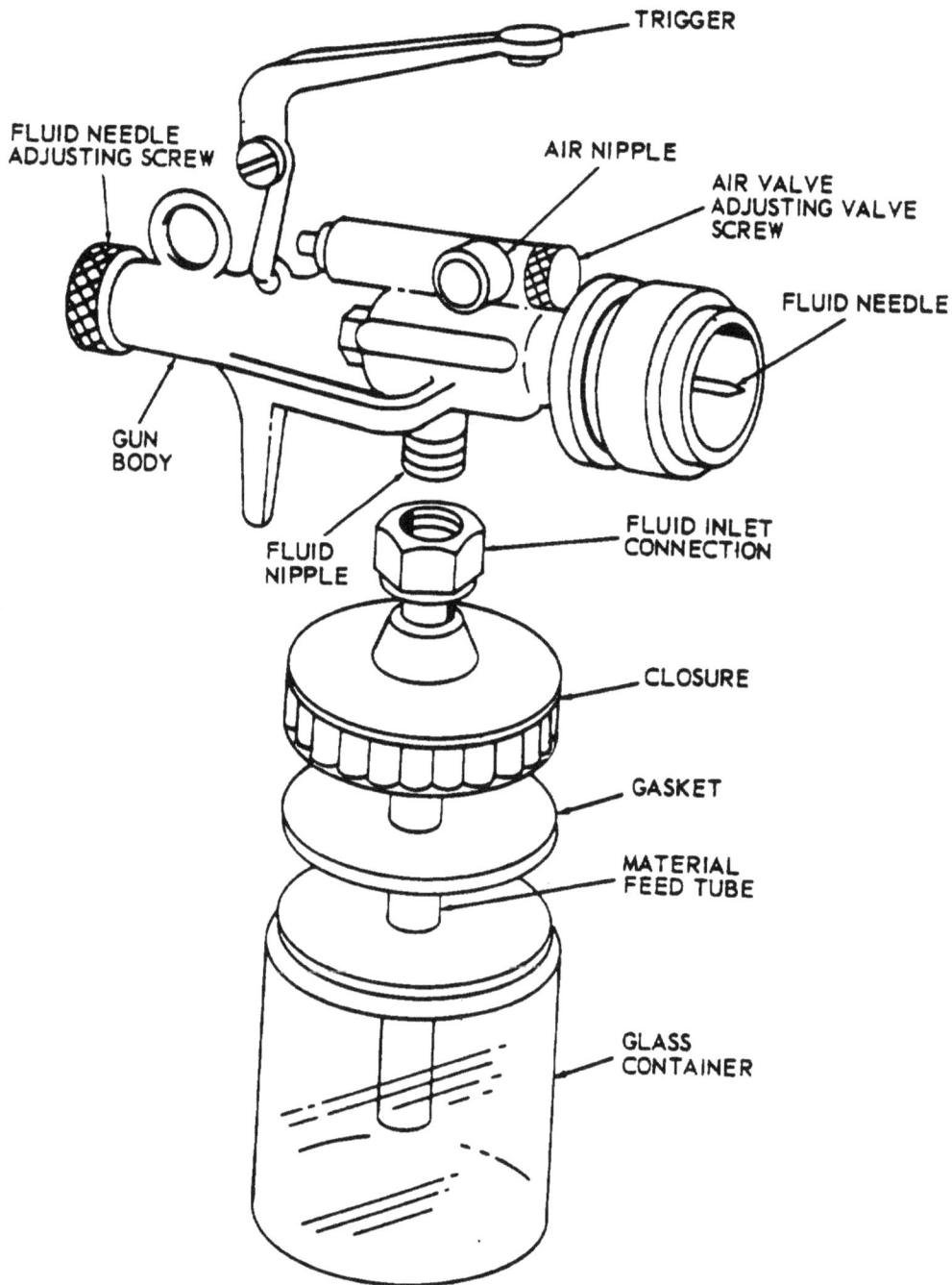

Figure 5-7. Attached-Container Type Spray Gun

e. Defective Spray Patterns

    (1) Heavy top pattern is due to:

        (a) Horn holes that are partially plugged.

        (b) An obstruction on top of the fluid tip.

        (c) Dirt on the air cap seat or fluid tip seat.

    (2) Heavy bottom pattern is due to:

        (a) Horn holes that are partially plugged.

        (b) An obstruction on the bottom side of the fluid tip.

        (c) Dirt on the air cap seat or fluid tip seat.

    (3) Heavy right side pattern is due to:

        (a) The right side of the horn holes is partially clogged.

        (b) Dirt on the right side of the fluid tip.

        (c) On a twin-jet cap, the right jet is clogged.

    (4) Heavy left side pattern is due to:

        (a) The left side of the horn holes is partially clogged.

        (b) Dirt on the left side of the fluid tip.

        (c) On a twin-jet cap, the left jet is clogged.

    (5) Heavy center pattern is due to:

        (a) The spray width of the adjusting valve is set too low.

        (b) The twin-jet cap, because the atomizing pressure is too low, or the material is of too great a viscosity.

        (c) The pressure-feed fluid pressure is too high for the air cap's capacity.

        (d) The nozzle is too large for the material being used.

    (6) Split spray pattern is due to an imbalance in the air and fluid pressure.

    (7) To remedy the conditions described in (1) through (4) above, determine if the obstruction is on the air cap or the fluid tip. Rotate the cap one-half turn and spray a test pattern. If the defect is inverted, the obstruction is on the air cap; if it is not inverted, the obstruction is on the fluid tip. Clean the air cap as instructed in paragraph 5-6. See figure 5-8 for examples of the conditions described in (1) through (4) above.

    (8) To remedy the conditions in (5) and (6) above, readjust the atomizing pressure, fluid pressure, and spray width adjustment until the desired spray is obtained. See figure 5-8 for examples of the conditions described in (5) and (6) above.

f. *Orange Peel Finish.* A common cause of this effect is the use of an improper or inferior thinner. With some thinners and paints, this condition may be noticed at certain times of the year due to atmospheric changes. Other causes are:

(1) Insufficient atomization.

(2) The gun is held too far from the surface.

(3) The gun is held too close to the surface, allowing air to ripple the surface.

(4) The material is not thoroughly dissolved or agitated.

(5) With synthetics and lacquers, drafts exist in the finishing room.

(6) With synthetics, there is low humidity.

(7) Improper (generally high) viscosity; the material should be reduced to specification requirements.

g. *Streaks in Finish.* Streaks are caused by:

(1) Tipping the gun, thereby causing one side of the pattern to deposit more material than the other. See figure 5-9.

(2) An improper spraying pattern.

h. *Sags and Runs in Finish.* Sags and runs are caused by:

(1) Tipping the gun, resulting in an uneven deposit of material.

(2) Too much material on the surface due to too much pressure, or gun travel that is too slow.

(3) Improper (generally high) viscosity; the material should be reduced to specification requirements.

i. *Mist or Fog.*

(1) This condition is caused by high atomization due to:

(a) An atomizing pressure that is too high.

(b) The wrong air cap for the material used.

(c) The wrong fluid tip for the material used.

(d) In pressure-feed systems, the fluid pressure is too low.

(2) It is also caused by improper use of the gun, specifically:

(a) Incorrect stroking.
(b) A gun that is held too far from the painting surface.

j. *Starving.* This condition is caused by insufficient air reaching the spray gun due to:

(1) The waste in the air transformer is packed too tightly or the air transformer is clogged with rust or dirt.

(2) The air cocks are too small.

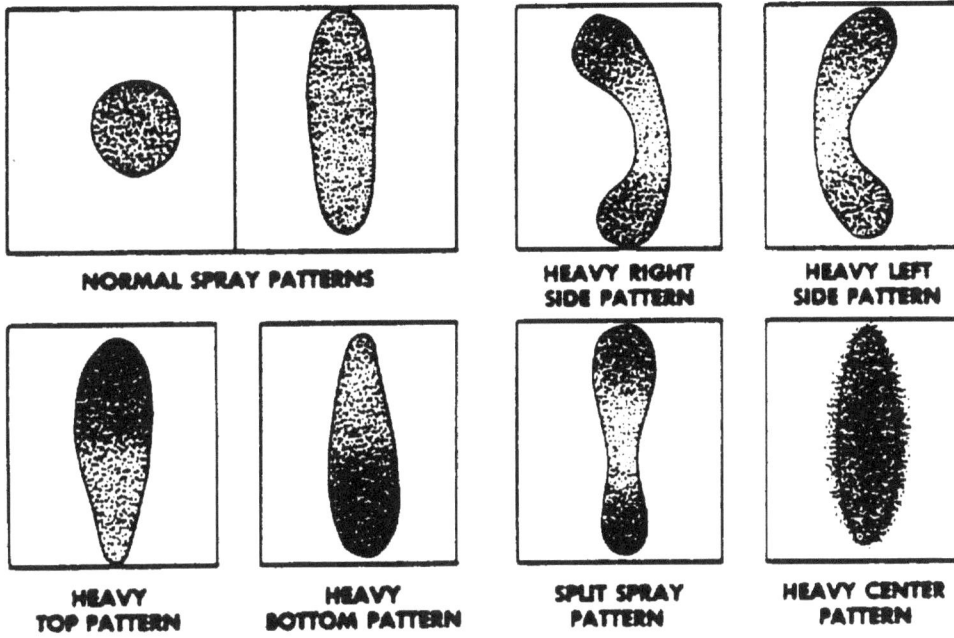

Figure 5-8. Defective Spray Patterns

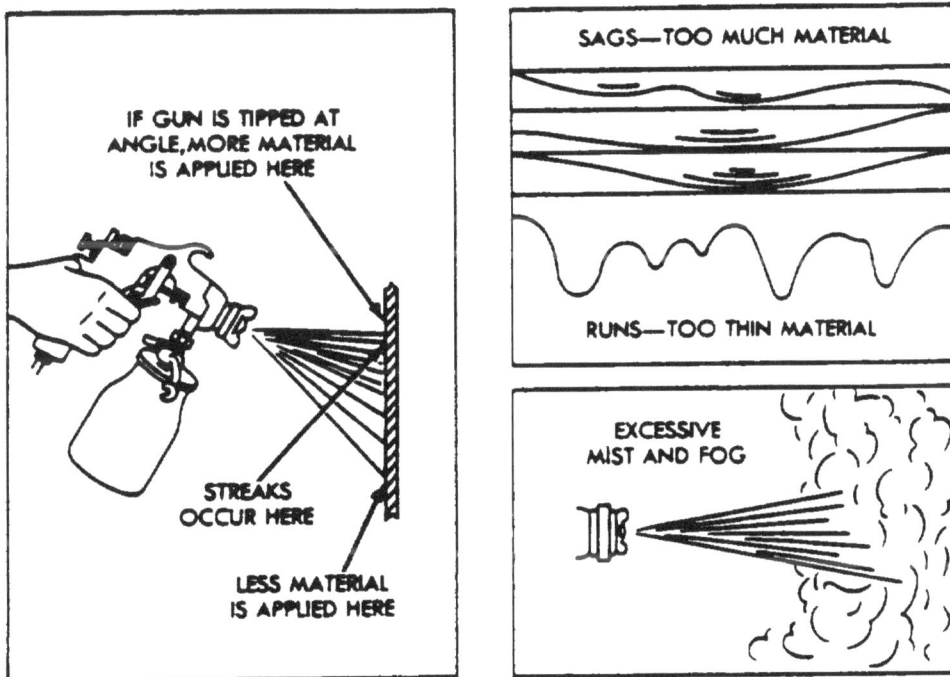

Figure 5-9. Spray Painting Faults

(3) The air lines are clogged.

(4) The air line is of an improper diameter.

(5) An inadequate air supply. This refers to the volume of air that is being delivered by the compressor, and not necessarily the pressure at which the air is delivered.

(6) The air intake is clogged.

5-6. CARE OF SPRAY EQUIPMENT

a. General. A spray gun is an instrument that has been designed and machined to close tolerances. Handle it with care so that the balance between the functional parts is not destroyed. Spray guns and related equipment require cleaning immediately after use. Paint that has hardened in a gun or hose is extremely difficult to remove, and usually causes a malfunction of the equipment. Be sure that the solvent used to clean the equipment is one in which the finishing material is soluble. Be sure to read the instructions that come with the pressure can regarding preservation of the nozzle.

b. Suction Cup Type. Remove the cup, keeping the fluid stem inside the cup or container, as shown in figure 5-10, then hold a cloth over the air cap and pull the trigger. This directs air into the fluid passageways, and blows any paint that may be in the gun back into the container. After cleaning out the cup, clean the gun by spraying a small amount of clean naphtha or thinner through it. Extreme care should be exercised in the removal of the fluid top so as not to split the tip or otherwise injure it or the fluid needle. When loosening the fluid tip, the trigger of the gun should be compressed so that the needle is not in contact with the tip; this eliminates the possibility of splitting the tip due to friction or sticking that is caused by dried paint. The entire spray gun should never be immersed in naphtha or thinner, as this removes lubricants and dries out the packing. Under no circumstances should the air or fluid ports of a gun or nozzle be reamed with any substance harder than soft wood, as a deformation of the spray pattern may result, and the spray gun may be rendered useless. To prevent wear, the working parts of guns should be kept lubricated with light machine oil. This requirement is especially true of the needle. The needle packing should also be kept pliable with an occasional drop of light lubricating oil.

c. Pressure Feed Type. Shut off the air supply to the pressure tank (see figure 5-11), release pressure in the tank, and blow back fluid as in the suction cup equipment (paragraph 5-6b above). Empty and clean the pressure tank. Place a container of clean naphtha or thinner inside the pressure tank and install the lid, making sure that the fluid delivery tubes (see figure 5-11) are immersed in the container of naphtha or thinner. Apply pressure and operate the spray gun to clean the hose and spray equipment. Disconnect the fluid hose from the gun and pressure tank and allow the hose to dry thoroughly before reconnecting. Clean the air cap and fluid tip as in the suction-cup equipment (paragraph 5-6b above).

5-7. TOUCHUP PAINTING

a. General. When equipment or material has spots from which the protecting paint has disappeared, and the rest of the paint surface is in a satisfactory condition, it is often advantageous to do a touchup, rather than a complete, painting job. The bare spots may have been caused by natural wear, abrasion, mechanical injury, or by rust or corrosion of the surface under the original paint. In such cases, it is necessary to clean the material beneath the spots and repaint using a method as near as possible to that used on the original paint job.

b. Cleaning. The spots to be painted must be thoroughly cleaned so that no decay, dirt, rust, corrosion, etc., remains. The remaining paint should also be worked down to a feather-edge if it is desirable to hide the lap.

c. Painting. While touchup painting may be done by the brush method, spraying is superior because the edges of the new paint can be feathered out to blend with the old surface. If the old and new colors match, the areas of the new paint will not be noticeable. In touchup work, use fillers, primers, and paints that are compatible with the existing undercoats, finish coats, etc. The general instructions for spray painting given in paragraphs 5-1 through 5-5 are also applicable to touchup work.

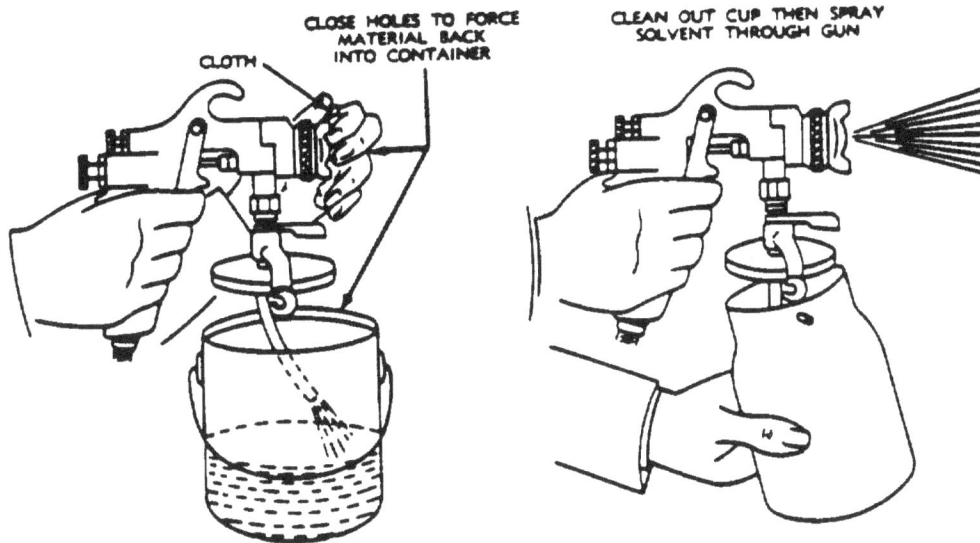

Figure 5-10. Spray Gun Cleaning

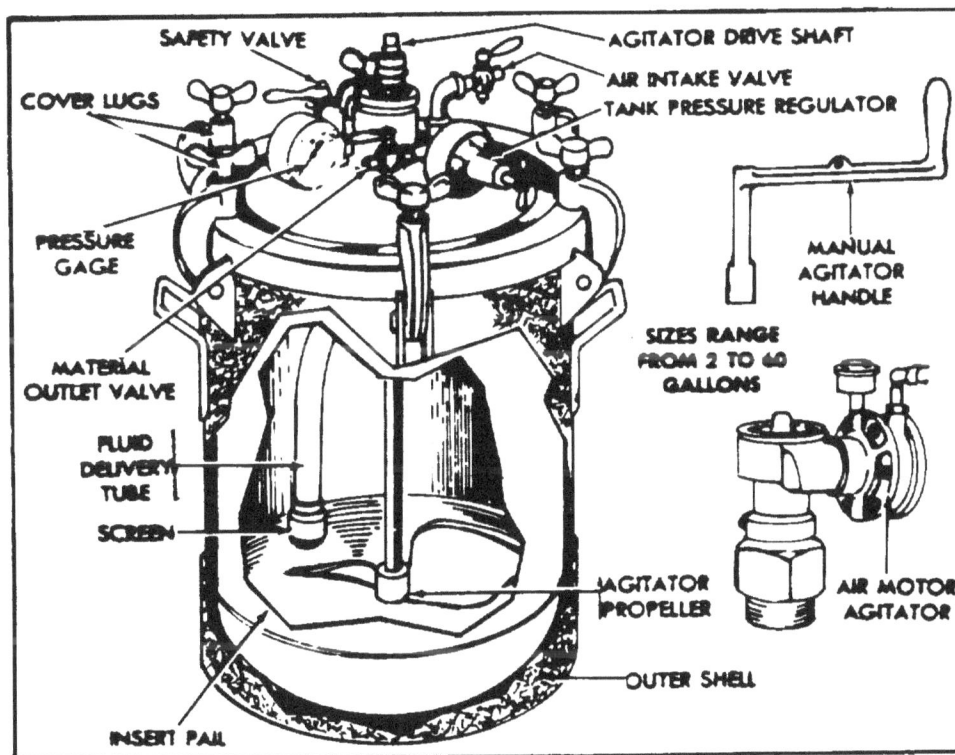

Figure 5-11. Pressure-Feed Paint Tank

5-8. ADDITIONAL SPRAY EQUIPMENT

 a. *Compressors.*

 (1) General. An air compressor is a mechanism designed to supply compressed air continuously at a predetermined pressure and volume. Compressors designated for spray painting are of two general types: single-stage and two-stage. These can be further divided into many other types, some of which are portable or stationary, unloader or pressure-switch controlled, have horizontally or vertically mounted tanks, are air or water cooled, and have a gasoline engine or an electric motor drive. Technical Manuals on air compressors are listed in AMC Pamphlet 750-9.

 (a) Single-stage compressors. A single-stage compressor (see figure 5-12) is one having one or two cylinders in which air is drawn from the atmosphere, compressed to a usable pressure, and delivered through an after cooler to the air receiver. Compressors of this type are intended for use where maximum pressures do not exceed 100 pounds per square inch (psi).

 (b) Two-stage compressors. A two-stage compressor (see figure 5-12) is one in which a relatively large cylinder first compresses the air to an intermediate pressure. Air compressed to this point is delivered through an intercooler to a small cylinder where it is compressed to the final pressure. It is then delivered through an after cooler to the air receiver. A system of this type is intended for use where required pressures exceed 100 psi. Such pressures will be encountered infrequently in everyday painting.

 (c) Gasoline engine drive. Gasoline engines of approximately three to five horsepower are used with compressors under the following conditions: where electric current is not available; where spray painting systems are used in localities served with different types of current; and in localities where insufficient current is supplied.

 (d) Electric motor drive. A majority of spray painting compressors are powered by electric motors of onefourth to five horsepower. The use of a system of this type is confined to locations in which the proper current is available. Electric motor-drive is generally chosen for more or less permanent installations, while a gasoline drive is preferred where portability is the prime consideration.

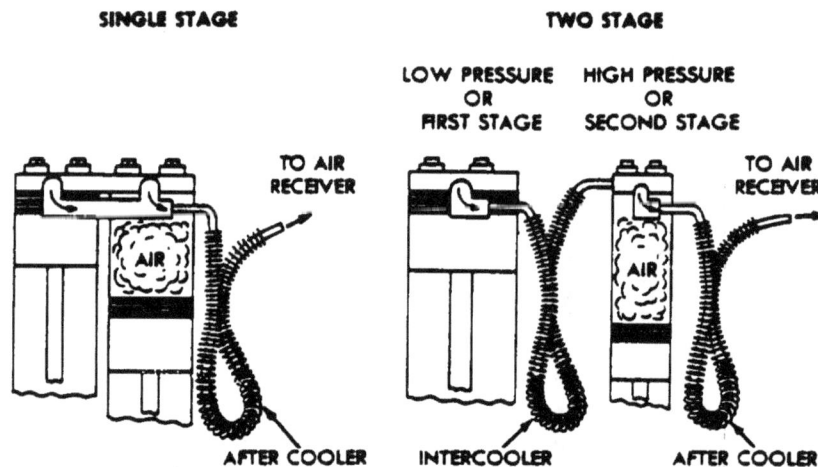

Figure 5-12. Single and Two-Stage Compressors

(e) Unloader and pressure switch control. Engine-drive compressors have unloaders which automatically disconnect the compressing cylinders from the air storage tank and allow the engine to continue to run at an idling speed until the tank pressure decreases to a preset minimum pressure. When the tank pressure reaches its preset minimum, the unloader valve automatically advances the engine throttle, opens the valves, and causes the pressure to be built up again. Electric motor-driven compressors have a pressure switch which shuts off the motor when a predetermined pressure has been established and restarts it when the pressure has fallen to a predetermined point.

(f) Horizontal and vertical tanks. Normally, compressor tanks are mounted in a horizontal position, serving as a base for the compressing unit. Where space is limited, or where ground clearance ora corner installation is important, tanks can be mounted in a vertical position.

(g) Air or water cooled compressors. The physical process of compression produces heat, and for that reason it is particularly important that air compressors be adequately cooled. Most compressors intended for use with spray painting equipment are air cooled. To accomplish this, the exterior surfaces of cylinders, intercoolers, and after-coolers are greatly increased in area by the use of fins. Increased surface area allows heat to be radiated more rapidly. Larger compressors, when used continuously, cannot be adequately cooled by air. Such compressors use a water cooling system consisting of a radiator, pump, fan, and water jackets built around the cylinders and are similar to those of automobile engines.

(h) Truck outfits. Air compressors, usually with gasoline engines for power, are often mounted on light, easily moved trucks. These portable units are ideal where painting is to be done at various locations.

(2) Inspection and lubrication of compressors.

(a) Gasoline engine.

NOTE

Do not make adjustments or repairs to gasoline engines unless qualified to do so.

Gasoline-driven air compressors are furnished with engines manufactured by many different companies. Generally speaking, they are of low horsepower and simple construction. For instructions on how to lubricate and adjust, refer to pertinent Technical Manuals. Adjustments to the carburetor, gasoline lines, and ignition systems should not be made in or near the paint shop. While the maintenance and repair of gasoline engines is not the responsibility of the painter, he should be sufficiently familiar with them to correct any minor stoppages caused by improper adjustment. Most gasoline engines used on compressors are adjusted at the factory to run at a constant speed and no throttle adjustments should be necessary.

(b) Electric motor. For lubrication of electric motors, refer to pertinent Technical Manuals. The adjustment of electric motors, even of a minor nature, is not the responsibility of the painter and should be undertaken only by qualified personnel. It is, however, the reponsibility of the operator to see that the electric motor on his compressor is not damaged through improper use. All electric motors, when overloaded, overheat, usually very rapidly. If this condition is allowed to persist for even a brief interval, the insulation on the winding may burn away. The use of fuses and circuit breakers is intended to prevent this, and the operator should familiarize himself with them. In no event should fuses be shorted or circuit breakers tied down.

(c) Care and preservation. For care and preservation of compressors, refer to pertinent Technical Manuals.

(3) Operation of compressors.

(a) Installation. Proper operation of a compressor system depends to a great extent on the correct initial installation of the equipment. The following points are important to proper installation: electrical wiring, whether for a permanent or portable installation, presents a fire hazard at all times, and should be installed and inspected by a competent electrician, as fire in a paint shop is devastating and almost impossible to control.

overload protection should be furnished for the electrical circuit; compressors should be located in an adjacent room and not in the paint shop, as this reduces the fire hazard considerably, improves the performance of the compressor, and reduces operator fatigue caused by the compressor noise; permanent installations should be at least one foot from adjacent walls to allow free air circulation over the cooling fins; air intakes should be piped to the outside of the building, where they can pick up clean, cool air; air pipe lines should be of sufficient size; an air compressor should be mounted on a solid foundation, because unless the weight is equally distributed, excessive vibrations will result in noisy operation and may cause a break in the tank supports or the compressing equipment; and the compressor should be installed so that it is level, with regard to a horizontal tank, and plumb, in the case of a vertically mounted tank, as this will assure the proper function of the compressor oiling system.

(b) Replacement. Like all mechanical devices, air compressors eventually wear out and should be replaced or rebuilt when: operational efficiency has decreased through wear and mechanical adjustments fail to restore it; or there is an insufficient air output which cannot be corrected by normal equipment adjusting; or the time interval from cut-in to cut-out is prolonged to the point of wasting power (a 50 percent deviation from the expected time interval is sufficient reason for replacing or rebuilding the system).

(c) Draining. The operation of compressing air, which always contains some moisture, induces condensation. Water condensed in this manner collects in the air receiver of the compressing outfit and must be removed each day by draining through the air receiver petcock. Proper location of the air intake will cut down the amount of water condensed in this manner.

(d) Servicing. Servicing of air compressors by personnel should be confined to the instructions given in this paragraph. Any further servicing by maintenance personnel should be performed in accordance with Technical Manuals on the specific compressors.

b. Tanks (Paint Containers).

(1) General. Material containers for spray painting systems are metal or glass vessels which are connected to the spray gun. These containers serve as supply reservoirs for the material to be sprayed, and are of a cup or tank type.

CAUTION

Painters planning to use coatings formulated with chlorinated solvents (such as MIL-C-46168, Type III) must remove all aluminum components from their painting system and replace with aluminum-free or stainless steel parts.

(2) Cup containers. Containers of this type are generally used where a variety of colors in rather small quantities are to be sprayed. There are two types: suction and pressure feed. Pressure-feed cups are recommended for small quantities of enamels, plastics, and other materials too heavy for suction feed, and where fine adjustment and speed of application are desired. The commercial preloaded and precharged pressure spray cans are unrechargeable.

(3) Tanks. Spray gun tanks are material containers for pressure-feed systems and provide a constant flow of paint at a uniform pressure. Their capacities range from two to 55 gallons. They consist of a container with a clamp-on lid, a fluid tube, outlet valves, a pressure gage, an agitator, and a safety valve. They also have an insertable paint container. They are furnished with either a top or bottom outlet and various accessories. There are two distinct types: Regulator type tanks offer the advantage of supplying large quantities of material to the gun under accurately controlled fluid pressure. Regulator-type tanks are frequently further subdivided into single and double regulator types, depending on whether the control is applied to the material pressure alone or to both the material and gun pressures. Regulator-type tanks are equipped with a pressure regulator, safety valve, release valve, etc., and are operated with different pressures on the air line and material. Equalized pressure tanks are equipped with only a safety and release valve, and operate with the same pressure on the air line and on the material.

(4) Insert containers. Insert containers are pail-like metal vessels designed to sit inside the tank. This eliminates cleaning the tank and facilitates the rapid change from one color to another. Their use permits several batches of material or colors to be mixed ahead of time. This construction also permits feeding directly from small cans of paint instead of from the full-sized container inside of the tank.

(5) Agitators. Certain materials require constant or frequent agitation while in the tank, and to meet this requirement, tanks are frequently supplied with mechanical agitators which can be activated by an air motor, by an electric drive, or by a manual crank.

(6) Material containers.

(a) General. Most metal containers are rugged, substantially built. and should present few, if any, operating difficulties. If regulator-type tanks are properly adjusted, air vents are kept free, and agitators are used when needed, little maintenance will be required beyond thorough and adequate cleaning procedures. Mounting a tank upon a dolly greatly extends the working area when used for multiple-gun operation. Tanks are available which provide for simultaneous two-gun operation.

(b) Precautions. Observing the following precautions will insure the proper operation of material containers:

WARNING

Never remove the cover from a pressure-feed container unless the pressure has been released.

Clean thoroughly after use, as many spray gun malfunctions can be traced to improper cleaning of the material container; be sure the fluid and air valve connections on the container are the proper size for the hose being used; test the tank safety valve regularly; keep the material containers full, as they do not function efficiently "I-' when nearly empty; and use the agitator regularly (where paint is being applied very rapidly, agitators are seldom needed, yet failure to agitate certain materials results in the formation of a surface skin which rapidly clogs filters and hose).

c. Hoses.

(1) Construction. Two types of hoses are used with spray guns: and fluid. An air hose has a red or orange cover while a fluid hose is black. The inner tube of a fluid hose is constructed of a solvent-resisting material that is generally impervious to any of the common solvents used in paint.

(2) Size. Hose of adequate inside diameter must be used with all spray gun systems. Too often a spray gun is blamed for improper paint feeding, or a material is considered of inferior quality, when the real cause of the trouble is low air pressure at the spray gun. Usually, this condition is caused by a hose that is too small. As seen in table 5-1, there is a natural pressure drop whenever compressed air is transmitted, and the amount of this pressure drop increases as the hose gets smaller.

(3) Pressure drop. Table 5-1 shows the air pressure drop expected from various lengths of one-fourth and five-sixteenths inch hose when used with a spray gun. For example, with 70 pounds of air pressure at the transformer, only 47 and one-half pounds of pressure (70 minus 22.5) will exist at the spray gun when 25 feet of one-fourth inch hose is used to connect the two units.

(4) Cleaning. The fluid hose should be cleaned immediately after use. In no event should a fluid hose be left uncleaned overnight.

(5) Storage. When not in use, the hose must be coiled and hung where it will be free from possible damage.

Table 5-1. Drop In Air Pressure

Air pressure at transformer (psi)	Air pressure drop at spray gun (psi)					
	5-foot length	10-foot length	15-foot length	20-foot length	25-foot length	50-foot length
air hose = 1/4 inch diameter						
40	6	8	9 1/2	11	12 3/4	24
50	7 1/2	10	12	14	16	28
60	9	12 1/2	14 1/2	16 3/4	19	31
70	10 3/4	14 1/2	17	19 1/2	22 1/2	34
80	12 1/4	16 12	19 1/2	22 1/2	25 1/2	37
90	14	18 34	22	25 1/4	29	39 1/2
air hose = 5/16 inch diameter						
40	2 1/4	2 3/4	3 1/4	3 1/2	4	8 1/2
50	3	3 1/2	4	4 1/2	5	10
60	3 3/4	4 1/2	5	5 1/2	6	11 1/2
70	4 1/2	5 1/4	6	6 3/4	7 1/4	13
80	5 1/2	6 1/4	7	8	8 3/4	14 1/2
90	6 1/2	7 1/2	8 1/2	9 1/2	9 1/2	16

d. *Valves and Gages.*

(1) Valves and gages used on spray painting equipment are of rugged construction and normally will need little attention to insure their correct operation. The following suggestions may be helpful in maintaining this equipment in good condition:

(2) Keep valves free from paint by wiping with a cloth dipped in solvent or thinner. Do not immerse valves in solvents or thinners, as this will dry out the packing.

(3) Be sure that valve nipples are of the correct size for the inside diameter of the hose being used. Incorrect mating of hose and nipple is a common cause of spray gun malfunction.

(4) Do not repair air gages in the field. Have this done by experienced and qualified personnel using the special tools and skills required for their proper adjustment.

e. *Air Transformers.*

(1) *General.* An air transformer, or separator, is a device which condenses oil and moisture, regulates and filters the air, and provides outlets to which spray guns and dusters may be connected (see figure 5-13).

(2) *Operation.*

(a) Oil and moisture are collected by the baffles and filter pack, allowing only clean, dry air to reach the spray gun. Further drying may be accomplished by the use of cartridges filled with a desiccant, such as silica gel, and installed in the outlets.

(b) Oil and moisture collect at the bottom of the air separator or transformer and should be removed daily (see figure 5-14).

(3) installation. Proper installation of the air transformer is essential to maintain correct operation. The following points are to be observed:

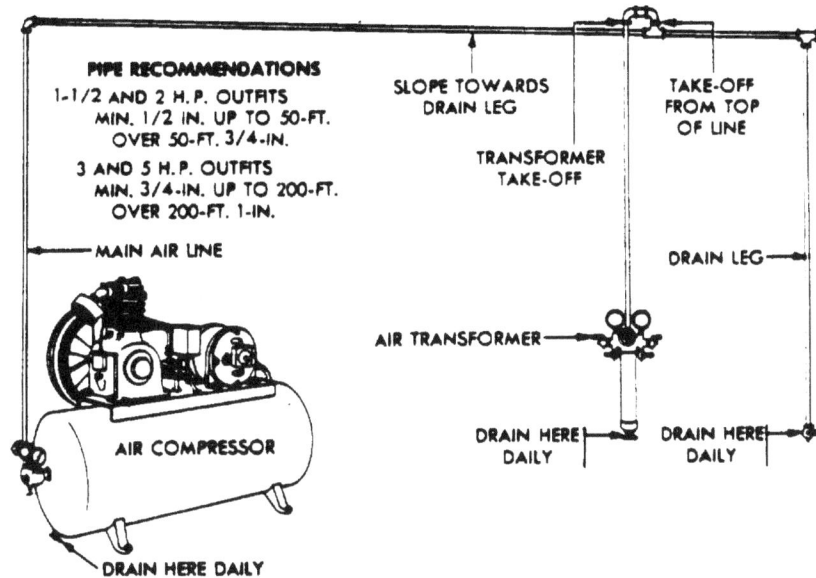

Figure 5-13. Air Transformer Installations

(a) Install the transformer at least 15 feet from the compressor.

(b) Air takeoffs from the compressor line to the transformer should be from the top of the line.

(c) The compressor air line to which the air transformer takeoff is attached should slant toward a permanently installed drain leg which should be drained daily. In localities where regulated air is available and only cleaning and filtering are needed, an air conditioner may be used to supplant the air transformer. The size of all necessary air lines is given in figure 5-13.

(4) Filter replacement. The filter pack in an air transformer should be inspected and replaced whenever it shows signs of becoming clogged by dirt or oil.

5-9. SHOP EQUIPMENT

a. Paint Booths.

(1) A paint booth is designed to collect, filter, and exhaust the fumes arising from the use of spray paint equipment. Due to Occupational Safety and Health Administration (OSHA), Environmental Protection Agency (EPA), and specific state and local requirements, any large scale (more than touchup) painting must be done in a paint booth. This is to minimize the release of toxic fumes into the environment and protect workers.

(2) Spray booths can be classified into two basic designs based on direction of airflow:

(a) Sidedraft Booths. Sidedraft booths have horizontal airflow. These booths take advantage of momentum of the spray mist and can be used when painting small to medium articles.

(b) *Downdraft booths.* Downdraft booths have vertical airflow. These booths permit greater protection while allowing more freedom of movement for the painter.

(3) Spray booths range in size from small, bench-type models to huge chambers capable of holding a large airplane. The basic consideration in determining the size of a spray paint booth is ensuring adequate space to permit the painters easy access to the top and sides of the object. If the object is transported by conveyor, the booth must be sufficiently long to allow coating within the time the object remains inside the booth.

(4) Booth exhaust air must be replaced for plant environmental control. The spray booth can be equipped with filter doors or fresh air inlets to reduce the amount of dust entering the booth. Air should enter the booth at low velocity (200 feet per minute (fpm) or less) and in the same direction as it is being exhausted to avoid unnecessary turbulence.

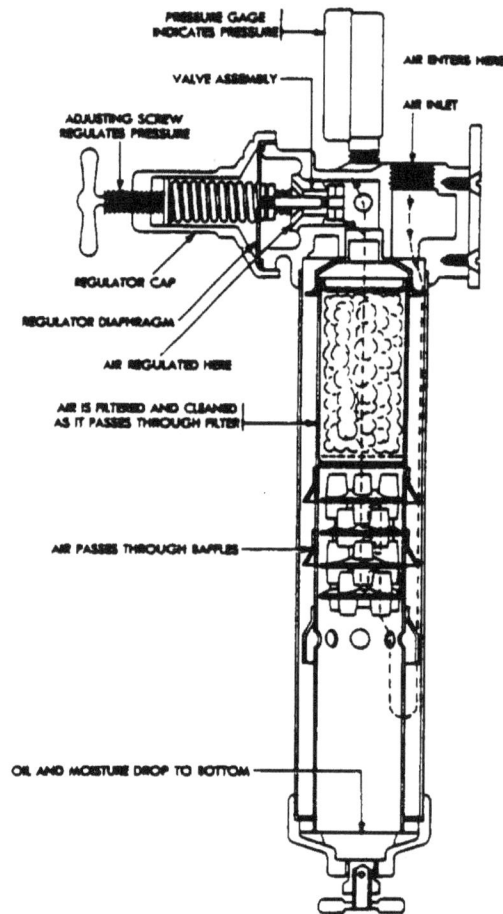

Figure 5-14. Operation of Air Transformer

(5) The booth air cleaning section not only removes paint mist from the exhausted air but acts as a means of air distribution within the booth. There are several types:

(a) Baffle type. An arrangement of metal baffles is simplest and provides a constant flow of air. Mist removal and clean-up difficulties limit its use to low production applications.

(b) Dry filter. These booths combine low cost with highly efficient paint mist removal, but have the disadvantage of a variable airflow. The airflow continuously decreases to a point where the filters require replacement. Dry filters must be disposed of carefully or a fire hazard will be created.

(c) Water wash. These incorporate various combinations of water curtains and sprays to scrub the paint mist from the exhausted air. They have the advantage of constant airflow, inherent fire protection, and high mist removal, but at a greater cost.

(6) Thought should be given as to how the booth will be maintained. Maintenance requirements can be reduced if:

(a) The booth is lined with strippable coating, such as wrapping paper or masking tape.

(b) Air filters are disposable.

(c) The glass shields over the booth lights are cleaned and coated with a light layer of white petroleum grease.

b. Portable Floodlights. Portable floodlights provide good illumination directly on areas to be painted. Their use will avoid uneven paint coverage. All lights used should be of the vaporproof type.

c. Can Shakers. A motor-driven paint can shaker (mixer) is a valuable piece of paint shop equipment. It saves considerable time and eliminates stirring by hand and paddle.

Section II. BRUSH OR ROLLER APPLICATION

5-10. WHEN TO USE BRUSH OR ROLLER APPLICATION

Brush or roller application is used when:

a. The volume of work does not justify setting up the spray apparatus.

b. Spray equipment is not available or is inaccessible to the job.

c. The operation is that of priming wood or other porous surfaces.

d. The task of masking-out non-painted parts is extensive.

e. Mist from the spray gun would damage the surroundings or create a fire hazard.

f. Respiratory equipment, required for spray application, is not available.

5-11. Selection of Brushes and Rollers

a. General. Factors determining the proper selection of a brush (see figure 5-15) for a specific task are:

(1) The material to be applied.

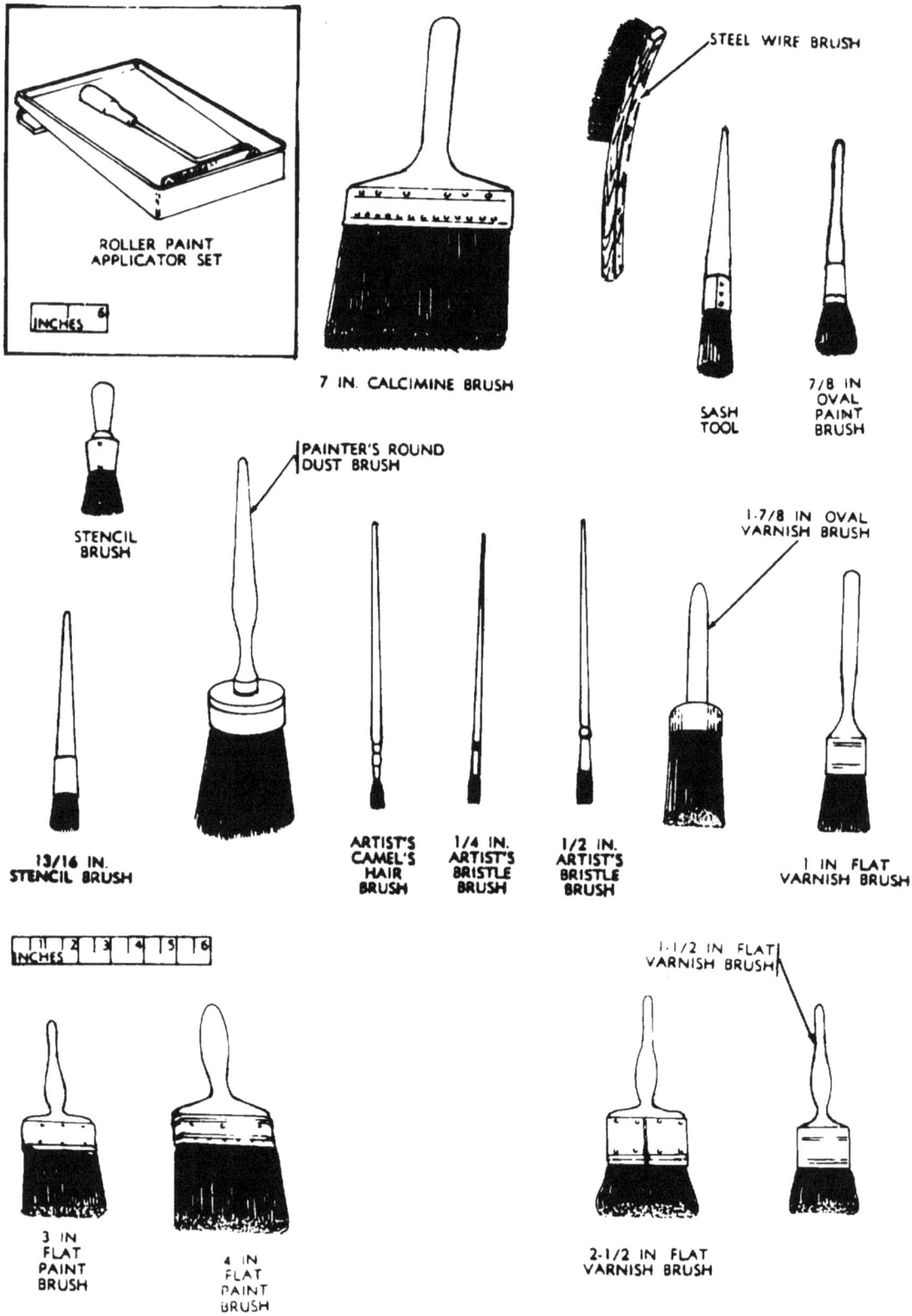

ROLLER PAINT
APPLICATOR SET

INCHES

7 IN. CALCIMINE BRUSH

STEEL WIRE BRUSH

SASH
TOOL

7/8 IN
OVAL
PAINT
BRUSH

STENCIL
BRUSH

PAINTER'S ROUND
DUST BRUSH

1-7/8 IN OVAL
VARNISH BRUSH

13/16 IN.
STENCIL BRUSH

ARTIST'S
CAMEL'S
HAIR
BRUSH

1/4 IN.
ARTIST'S
BRISTLE
BRUSH

1/2 IN.
ARTIST'S
BRISTLE
BRUSH

1 IN FLAT
VARNISH BRUSH

INCHES 1 2 3 4 5 6

1-1/2 IN FLAT
VARNISH BRUSH

3 IN
FLAT
PAINT
BRUSH

4 IN
FLAT
PAINT
BRUSH

2-1/2 IN FLAT
VARNISH BRUSH

Figure 5-15. Types of Brushes and Rollers

(2) The nature of the surface on which the material is to be applied.

(3) The area to be covered.

b. For Stains. Brushes with rather stiff bristles, preferably set in rubber, are used to apply stains on wood with open pores. The stiffness of the bristles is essential in working the stain into the pores of the wood. A softer brush is needed for close-grained wood.

c. For Paints. Flat brushes with long soft bristles or hair are required to apply paint. The width will vary with the area and nature of the surface to be covered.

d. For Enamels. Brushes used for enamels should be relatively large, with a chisel point. Skunk hair (fitch), rubber-set, varnish brushes with moderately soft and fine bristles are best. This type of brush, slightly moistened with water, can be used to touch up a non-CARC surface by rebrushing the coat, providing the brushing is done soon after the enamel film has been applied.

e. Paint Rollers. Paint rollers are replacing brushes more and more. There are three types: quench, fill, and power. The quench roller requires a tray for quenching with paint. The fill roller does not require a tray, but a funnel is needed for filling. Both rollers require buckets and/or trays for easy handling. The power roller has an electric motor that pumps paint directly from the container to the roller. These rollers are used mostly on large wall areas, floors, and ceilings. The material to be applied and the nature of the surface to be treated are factors that will govern their usefulness.

5-12. BRUSHING VARIOUS MATERIALS

a. General. Brushing is used where rolling is impractical. A right-handed operator should start at the right edge of the surface to be painted and proceed toward the left. Using this procedure, the full paint brush is applied to the uncoated surface by brushing back into the wet film. By decreasing pressure at the end of a stroke, brush marking is minimized. A left-handed operator should start at the left edge of the surface to be painted and proceed toward the right.

b. Stain. Apply freely, rapidly, and evenly in the direction of the wood grain, and brush well into the pores. Certain types of stain must be wiped off with clean, lint-free cloths in order to produce a uniform effect.

c. Paint. On exterior woodwork, use a long, sweeping, straight stroke.

d. Slow Drying Enamel. Make short strokes in one direction until a small area is covered, then go back over the area with strokes at right angles to the first, in order to obtain a smooth even coat. This operation is called "laying-off" the finish. Follow with an adjacent area of similar size before the enamel in the first area sets.

e. Lacquer, Quick Drying Enamel, and Shellac. These materials must be applied rapidly. Each stroke of the brush must completely cover the area to be traversed, and the brush must be kept well-charged with material so that no retouching of spots is required; retouching results in a rough finish. This procedure is sometimes referred to as "flowing on" a coat.

f. Varnish. Use a well-charged brush, and depending upon the speed with which the varnish dries, quickly "flow on" the coat. If possible, "lay-off" the finish to give a smooth film.

g. Brushing Technique. See figure 5-16.

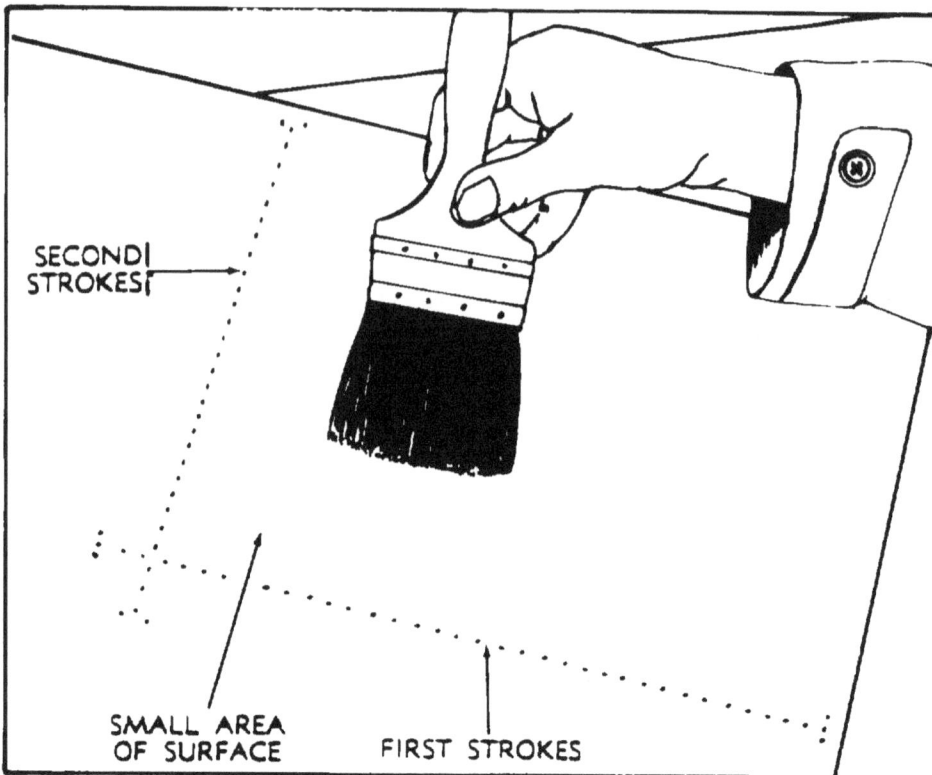

A. BRUSHING TECHNIQUE FOR VARNISH AND ENAMEL.

B. TECHNIQUE FOR PAINTING A CEILING.

Figure 5-16. Brushing Techniques (Sheet 1 of 2)

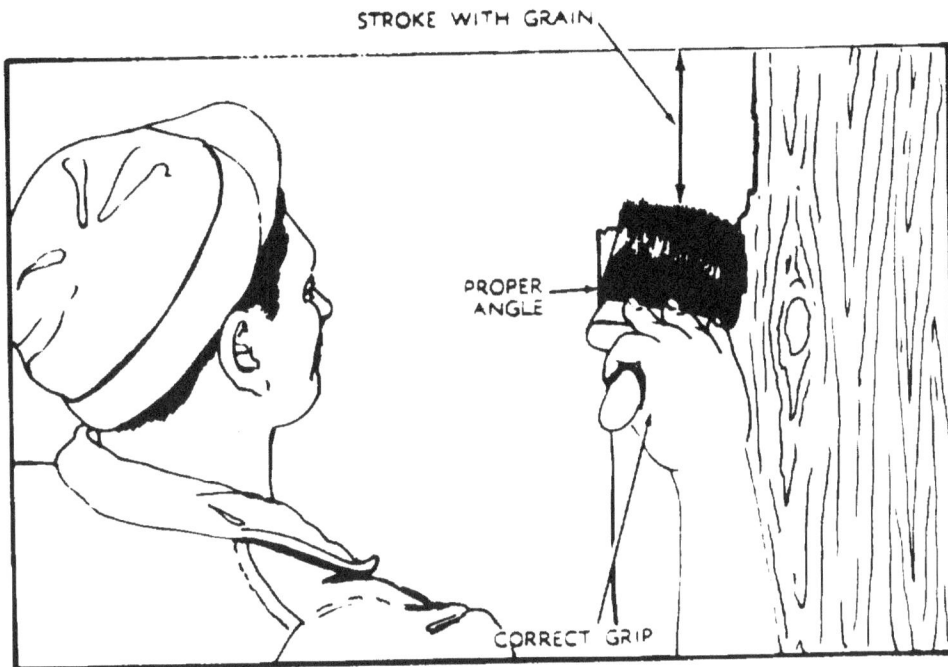

C. BRUSH TECHNIQUE FOR FLAT SURFACES.

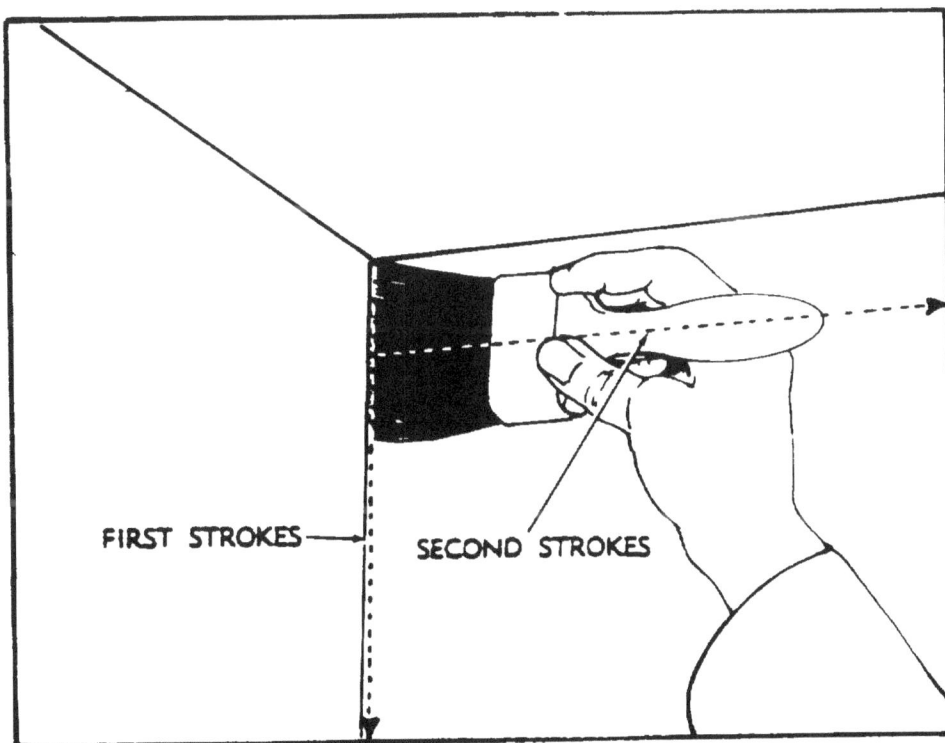

STROKE WITH GRAIN

PROPER ANGLE

CORRECT GRIP

FIRST STROKES

SECOND STROKES

D. BRUSH TECHNIQUE FOR PAINTING IN CORNERS.

Figure 5-16. Brushing Techniques (Sheet 2 of 2)

5-13. CLEANING OF BRUSHES AND ROLLERS

a. General. In order to keep paint brushes soft and pliable, they should be cleaned immediately after use. Once the material has been allowed to stand overnight, no amount of cleaning will restore the original pliability or remove the hardened material from the heel of the brush. Solvents or thinners used with the material just applied by the brush are the best possible cleaners. Mineral spirits paint thinner, xylene (ASTM D846), turpentine, and synthetic enamel thinner are some of the common brush-cleaning fluids.

WARNING
Xylene is a hazardous material and must be disposed of in accordance with AR 420-47, Solid and Hazardous Waste Management.

b. Steps in Cleaning.

(1) Save cleaning materials by first pressing the brush firmly against the rim or side of the paint container, thus squeezing out as much paint as possible.

(2) In the event the paint has hardened in the brush, it should be softened and carefully worked out with a putty knife.

(3) Pour a small amount of brush cleaning fluid into a shallow, wide-mouthed container and work it thoroughly into the brush, making sure that the fluid gets up to base of the bristles.

(4) When this small amount of fluid is loaded with paint from the brush, discard the fluid. Take a somewhat larger amount of fresh fluid and repeat the operation as many times as necessary until all traces of pigment and paint disappear and the bristles are soft and pliable.

(5) Paint rollers are cleaned in much the same manner. Use a pan instead of a pail and soften by following the instructions given by the roller manufacturer.

5-14. CARE AND STORAGE OF BRUSHES AND ROLLERS

a. General. The proper care of paint brushes and rollers requires the use of a few basic rules:

(1) Never stand brushes, wet or dry, on their bristles. This will cause the bristles to become permanently bent or deformed and will ruin the brush.

(2) A brush used periodically should be stored in a keeper, such as a container of linseed oil or another appropriate thinner. Suspend the brush from a nail or hook so that the bristles are covered with thinner but are not touching the bottom of the container.

(3) Brushes that are not frequently used should be thoroughly cleaned with the proper paint thinner or cleaning agent. After drying, they should be stored in a wrapper to retain their shape.

(4) For care of rollers, follow the manufacturer's instructions.

b. Storage Overnight.

(1) Paint brushes in daily use should be kept overnight in a brush keeper. Immersion of the cleaned brush in oil or thinner will assure that the bristles will remain soft and pliable. Segregate brushes in their keepers according to the type of material used. Use a MIL-T-81772 keeper for brushes used with CARC; use a linseed oil keeper for brushes that are for use with paints and varnishes; place dope and lacquer thinner in the keeper for brushes used with lacquers; use synthetic-resin enamel thinner for brushes that are used with enamels; and use alcohol, MIL-STD-1201 or O-E-760, for brushes that are used with shellacs. A keeper cover should be used to prevent evaporation and contain the flammable vapors of solvents and thinners.

If necessary, drill a hole in the brush handles for suspension in the keepers.

(2) Use enough oil, solvent, or thinner in the keeper so that the bristles of the brushes are covered. These brushes should not touch each other or the bottom of the container. Brushes kept in linseed oil should be cleaned before use by washing in thinner.

NOTE
Brushes used for CARC, lacquer, synthetics, or shellac should be placed in brush keepers containing CARC thinner, lacquer thinner, synthetic thinner, or alcohol, respectively. Traces of linseed oil will spoil such materials and the finishing job performed with these brushes.

(3) For rollers, follow the manufacturer's instructions.

c. *Indefinite Storage.* When brushes are not to be used for a long time, they may be prepared for storage as follows:

(1) Clean thoroughly.
(2) Immerse in raw linseed oil or another appropriate thinner for a few days. This can be done in the brush keeper.
(3) Remove from the keeper and press out most of the thinner.
(4) Straighten the bristles and wrap the brush in paper. Brushes treated in this manner should be stored flat with no weight applied to the bristles. Open the package and repeat the procedure every six months or less.
(5) A procedure similar to this should be followed for storage of rollers. Follow the instructions given by the manufacturer for these items.

5-15. MISCELLANEOUS EQUIPMENT

a. *Scrapers.* Scrapers of various sizes, made of bronze, which do not produce sparks when rubbed on other metals or concrete surfaces, are used for cleaning paint residue from spray booths, the floor, and from metal and wood surfaces. Flexible carbon scrapers should be used on aluminum and magnesium since metal scrapers may leave deposits of metal. These deposits promote galvanic corrosion, and in some instances, could even cause shorting of electrical circuits.

b. *Stencil Sets.*

(1) *Brass stencils.* Brass stencils in one to four inch sizes are sometimes used for stenciling.

(2) *Paper stencils.*

(a) Star-insignia cardboard stencils are available in various diameters.

(b) Paper stencil sets are available in various sizes.

(3) *Gummed-back paper stencils.* Gummed-back paper stencils are available for applying registration numbers.

c. *Miscellaneous Tools and Supplies.*

(1) *Layout Tools.* Straightedges, a yardstick, a steel square, and dividers are used for laying out lines to guide in the location of letters and insignia when stenciling. Guidelines can also be made by snapping chalked string against the surface to be painted.

(2) *Hydraulic jack.* A good hydraulic jack is required for the removal of vehicle wheels before painting. Wooden or iron horses are sometimes needed for this operation to support the vehicle with its wheels removed. Inspect jacks prior to each use to ensure that they are safe for use (i.e. no leaks, cracks, etc.).

(3) *Supplies for preparing surfaces*. Painting, removing, cleaning, rust-removing solutions, and sanding materials are required for preparing surfaces.

(4) *Masking tape.* Masking tape is required to cover all body parts that are to be protected from paint spray. Tape alone is used to mask small areas. For larger areas, such as windows, the tape is used to fasten paper over the area to be protected.

(5) *Sanding disks.* Sanding disks are used with a motor sander and polishing pads and solutions are used with an electric buffer.

(6) *Other tools.* Other tools required in the paint shop include paint brushes, wire brushes for cleaning off loose paint and rust, and putty knives or scrapers for removing old paint. Razor blade scrapers are useful for removing paint from glass. A 16 ounce graduated glass container is needed for mixing paint and thinners in the required proportions.

(7) *Cloths.* An abundance of wiping cloths is required for wiping off spilled paint and for cleaning spray guns and related equipment.

WARNING
Do not use electric sanders in a paint shop or near a spray paint area.

d. *Electric Sanders.* Portable, motor-driven, disk or orbital sanders are occasionally required for smoothing a vehicle's body or fender before it is painted, although this is not usually the work of the paint shop.

e. *Electric Buffers.* Ordinarily, the same tool is not used for both sanding and buffing because the sander rotates much faster than the buffer. There are combination sanding/buffers, however, that run at different speeds to accommodate both operations.

Section III. DIP APPLICATION

5-16. WHEN TO USE DIP APPLICATION

The dipping method of applying paint is generally used for small articles and is especially suited to the coating of items of irregular design that are difficult to reach by brush or spray; for example, the interior of a narrow tube. Dipping is not time or cost effective except when a large number of items are to be painted in a production line manner. CARC primers and coatings should not be used for dipping.

5-17. DIPPING TECHNIQUES

Ensure that the paint has been thinned to dipping consistency. Suitable consistencies vary with each article, and must be arrived at by trial and error. Suspend the article with a cord or wire and immerse in paint. Remove the article slowly, hang from a line, and allow it to dry in a comparatively draft-free location over a dipping tank or draining pan.

5-18. EQUIPMENT REQUIRED

A receptacle to hold the paint is required. This can be a pail or a specially constructed tank. In general, the receptacle should be just large enough to conveniently permit the insertion of the article to be coated. Replenish the paint as needed and use paddles to stir at frequent intervals. If the receptacle is large, a drain-off valve should be provided so that the paint may be removed and placed in sealed containers when the dipping operations are completed.

Section IV. TROUBLESHOOTING TECHNIQUES

5-19. GENERAL PAINT FAILURES

a. There is a cause for every paint failure, and in most instances, the failure can be prevented by observing specific precautions and instructions. The weather, with its humidity, heat, cold, sudden rainstorms, etc., can damage a paint film, and for this reason the painter should take into consideration the atmospheric conditions prior to painting.

CAUTION

Coatings should not be applied at temperatures below 500°F (10°C).

b. The most frequent causes of paint failure are discussed in paragraphs 5-20 through 5-34.

5-20. ALLIGATORING AND CHECKING

a. Characteristics. When a rupturing of the top paint coat causes the surface to break up into irregular areas (separated by wide cracks in alligator-hide style), the condition is referred to as "alligatoring" or "checking." Alligatoring on a painted surface can be detected by the appearance in the top coat of small openings or ruptures which divide the surface into small irregular areas, leaving the undercoat visible through the breaks in the top coat.

b. Probable Cause. Alligatoring is usually caused by too soft an undercoat or by applying a coat over an underlying coat which has not thoroughly dried.

c. Corrective Measures. Remove the entire paint coat using a scraper or paint remover. Mild cases should be thoroughly wire brushed. Before repainting, clean the surface after the old paint has been removed.

5-21. BLEEDING

a. Characteristics. When the color of a previous coat is absorbed into the topcoat, the condition is called "bleeding."

b. Probable Causes. Bleeding is usually caused by the partial solution of the old pigment into the new coat. Bleeding may also occur with asphalt and colored resins.

*c. **Corrective Measures**.* The corrective measures to be taken depend on the severity of the bleeding and the quality of the appearance required. If bleeding is not severe, and appearance is not important, apply another coat of paint after the previous coat (in which bleeding occurred) has dried thoroughly. If this method fails to provide an acceptable finish, remove all paint coatings, clean the surface thoroughly, and repaint.

5-22. BLISTERING

a. Characteristics. Blistering is evidenced by blister-like irregularities on the film of a painted surface, with the paint coat detached and raised from the surface upon which it is applied.

b. Probable Causes. Blistering is the result of gases or liquids (usually water) forming under the coating. The most common cause of blistering on wood surfaces is the application of paint over a damp or wet surface. The breaking of the blisters may result in a peeling of the paint coat. Blistering is also caused by using a paint that is incompatible with that used in a previous coating.

c. Corrective Measures. Use a wire brush or scraper to remove all defective paint. Permit the surface to dry thoroughly, then repaint.

5-23. BLUSHING

a. Characteristics. A surface on which blushing has occurred is characterized by a white discoloration in the coating and sometimes by the separation of ingredients from the coating. Blushing most commonly occurs in nitrocellulose lacquers.

b. Probable Causes. Blushing may be caused by condensation of moisture on the film or by improper composition of the vehicle (pigment-carrying liquid portion of paint) or solvent.

c. Corrective Measures. Remove or sand the film where blushing has occurred and repaint (after insuring that the surfaces are dry). Blushing on acrylic lacquer may be prevented by adding acrylic lacquer retarder to the liquid lacquer.

5-24. CHALKING

a. Characteristics. Chalking can be detected by the existence of dry, loose powder on the paint film. Rain tends to wash this powder off of exterior surfaces.

b. Probable Causes. The chalking of a painted surface is governed partially by the composition of the paint. Chalking, loss of luster, and deterioration of the surface film are also affected by atmospheric conditions. Paints low in binder content, or high in inert pigments, have a tendency toward early and excessive chalking.

c. Corrective Measures. A paint which chalks moderately affords a better repainting surface than one which does not chalk at all; however, if excessive chalking has taken place, remove all the loose and powdery substance from the surface with a wire brush and repaint.

5-25. CRACKING, FLAKING, SCALING, AND PEELING

a. Characteristics. Breaks which extend through the paint film to the bottom surface are called cracks. Cracking is usually followed by flaking, scaling, or peeling. Flaking is the dropping off of small pieces of the paint coat. Scaling is an advanced form of flaking and is evidenced by larger flakes. Peeling is the curling and dropping off of relatively large pieces of paint film.

b. Probable Causes. Paints which become brittle when dried cannot contract or expand with moisture and temperature changes, and are very susceptible to cracking. Cracking may also be caused by too many coats being built up due to previous painting. Cracking advances to scaling and peeling as the old paint, which has lost its elasticity and much of its adhesive grip, is pulled loose by the surface tension of the new paint film as it dries. Low grade paints usually lack elasticity because they are deficient in oil and contain too much inert material for extended exposure. Since flaking and scaling are usually preceded by cracking, their causes are much the same as for cracking. Scaling and peeling frequently occur when paint has been applied to unseasoned or damp lumber. Peeling may also occur around knots, and where cracks in the paint permit water to get behind the paint film.

c. Corrective Measures. Use a wire brush to remove all loose paint. In the case of cracking, remove the entire paint coat using a scraper or paint remover. Clean the surface thoroughly with a duster before repainting, and be sure that the first coat is thoroughly dry before applying a second coat.

5-26. CRAWLING OR CREEPING

a. *Characteristics.* Crawling or creeping of paint is noted by little drops (or islands) which form on the paint film.

b. *Probable Cause.* Crawling often occurs when varnish or enamel is applied on an oily or greasy surface. Painting over a very smooth surface will sometimes cause crawling.

c. *Corrective Measures.* Remove the little islands of paint which have formed on the film by sanding them, and wash off any grease or oil which may be underneath. If a glossy coat has been applied over another glossy coat, remove both coats using varnish and paint remover. Apply a prime coat without gloss before applying a high gloss topcoat.

5-27. DULLING

a. *Characteristics.* Dulling is characterized by the loss of gloss which should be present in a high gloss varnish, paint, or enamel film after it has dried.

b. *Probable Cause.* Dulling may be caused by the action of gases, inferior products, use of very old stock, or the use of too much turpentine or thinner.

c. *Corrective Measures.* Remove the dulled coat, or sand it down with fine sandpaper, and apply a varnish, paint or enamel of known good quality.

5-28. MILDEWING

a. *Characteristics.* Mildew is a fungus frequently found on exposed surfaces in warm, damp climates, particularly on soft paint films.

b. *Probable Causes.* Paint film that has become sticky or tacky attracts windblown spores and decayed and dried vegetation to its surface. The oil in the paint sometimes becomes infested, and the breeding of mildew spores takes place.

c. *Corrective Measures.* To prevent the recurrence of mildew, the old coat of paint should be removed and a new coat of hard-drying paint applied. A fungus growth can be partially removed by scrubbing the affected surface with a solution of trisodium phosphate and water. The surface should then be rinsed with clear water and allowed to dry. The use of less paint and more thinner is advised in environments where mildew is a common occurrence.

WARNING

Extreme care must be observed in the handling of paints containing mercury or other fungicides to prevent poisoning or skin irritation.

5-29. STREAKING AND LUMPING

a. *Characteristics.* Streaks or lumps on painted wood surfaces are caused by resin and pitch exuding from knots and unseasoned lumber.

b. *Probable Causes.* This condition is caused by painting over unseasoned lumber and by painting over knots or resinous streaks which have not been properly treated before painting. On metal, it is an indication that the paint has been applied incorrectly.

c. Corrective Measures. Apply shellac, varnish, or aluminum paint to wood knots before painting. Do not paint unseasoned wood. For metal, apply paint with a spray gun, holding the gun level so that an even coat is applied.

5-30. RUNNING AND SAGGING

a. Characteristics. An effect of ripples or irregularities in a film of paint, varnish, or lacquer is known as runs or sags.

b. Probable Causes. Runs or sags are usually produced by the application of a paint, varnish, or lacquer which has been thinned excessively, or by the application of too much material. It is usually evident on a sloping or vertical surface. Other causes are incomplete brushing or the use of a stiff brush.

c. Corrective Measures. Sand the surface until runs or sags have been removed, then recoat with material of the correct consistency, taking care not to apply excessive amounts. Use a flexible brush for this operation.

5-31. SLOW DRYING

a. Characteristics. Although the time required for drying is dependent upon the type of paint, enamel, varnish, or lacquer used, certain weather conditions may prolong the drying period. Paints which, under normal drying conditions, are tacky or sticky for long periods (12 hours or longer) are likely to attract dust and dirt, to promote mildew, or to develop checking or alligatoring.

b. Probable Causes. Cold weather retards drying. Drying agents also may lose their effectiveness in prepared paints that are dark in color. The use of old thinner, or the use of inferior driers and thinners, are other factors frequently contributing to slow drying of paint films.

c. Corrective Measures. Do not paint when the temperature is below 50°F (10°C). A standard procedure is to paint a test area and let it dry overnight before adding additional drier to the paint. This is done to assure a correct drying period. In cold weather, apply a thin uniform film on a dry surface.

5-32. SPOTTING

a. Characteristics. The appearance of discolored spots on a painted surface is known as spotting.

b. Probable Causes. Color changes and loss of gloss in irregular patches may be caused by spots in the surface which absorb oil from the paint unevenly. This may be the result of too few coats, or the lack of controlled penetration of the paint, and may occur on new items which have been given only two coats, or an old item painted with just one coat. In white paints, this is accompanied by the loss of gloss. Colored paints usually appear to fade when the oil is absorbed unevenly. Spots are sometimes caused by nail heads which rust. Splashes of liquid on a freshly varnished surface will cause spotting, and rain or hail on a freshly painted surface will also leave spots.

c. Corrective Measures. Apply an additional coat of paint. Apply paint during dry weather. The use of paint containing zinc oxide is effective in minimizing spotting on older items. In cases of spotting due to rain or sandstorms, sand off rough spots before repainting.

5-33. SWEATING

a. Characteristics. The reappearance of luster on a varnished surface which has been rubbed to a dull finish is known as sweating.

b. Probable Causes. Sweating of a varnished surface is usually caused by inadequate rubbing to attain a dull finish, or the application of a finish coat before the undercoats have thoroughly hardened.

c. Corrective Measures. After the surface is thoroughly hardened, rub down the finish thoroughly and then apply another finish coat.

5-34. WRINKLING

a. Characteristics. Wrinkling of a paint coat is evidenced by the paint film gathering in small wrinkles.

b. Probable Causes. Wrinkling may be caused by the application of an excessively thick coat, or by a failure to brush out the paint properly. Wrinkling may also be caused by too much drier in the paint. Paints which have been excessively thinned with oil and applied thickly are also subject to wrinkling.

c. Corrective Measures. Sand off the wrinkles with rough sandpaper and paint with properly thinned paint which does not have an excessive amount of drier or oil in it. In cases of excessive wrinkling, strip off the old coats and repaint. Wrinkling in acrylic lacquer may be prevented by adding acrylic lacquer retarder.

(5-35/5-36 Blank)

CHAPTER 6

MARKING PROCEDURES

Section I. LETTERING AND SIGN PAINTING

WARNING

Before beginning any painting-related activity, read Chapter 1, Section II, Safety Summary.

6-1. PURPOSE

The directions given in this section are designed to acquaint the painter with the basic principles of lettering and sign painting.

6-2. LETTERING STYLE

The lettering style suitable for all military requirements is known as the Vertical Gothic Style, illustrated in figure 6-1. There will be times when stencils and decals are not available, requiring the soldier to utilize hand lettering. The types of brushes required, and lettering and painting techniques used, are described in the following paragraphs.

6-3. BRUSHES

   a. *Rough Surfaces.* Painting brick, concrete, stucco, rough plaster, and boards that have been painted before will require a flat bristle brush. The size depends upon the width of the letter. Brushes for these surfaces are classified as fitches, angular fitches, and cutters.

   b. *Smooth Surfaces.* Painting metal, glass, vehicles, boards, hardboard, and cardboard, will require a softer, flat bristle, artist-type brush, or an oval wash brush, to obtain a finer degree of finish. These brushes are classified as single stroke, lettering brushes, and come in a variety of soft bristle combinations. For beginner's use, a flat oxhair-and-sable combination is suggested. This type of lettering brush has a knifelike precision edge and will hold a large load of paint, which feeds evenly and accurately to the surface; it is also easy to control.

6-4. LETTERING TECHNIQUE

   a. *Preparation of the Brush.* Dip the brush into the paint until all the bristles are immersed. Raise the brush straight up until all excess paint drips from it. Stroke the brush back and forth on a smooth, flat surface in razor strop style until the paint is worked well up into the bristles and until the end of the bristles form a sharp chisel-like edge (see figure 6-2). This makes it possible to form a sharp, uniform stroke.

   b. *Basic Strokes.* For lettering, an oval wash brush should be used because of its rounded end. Three basic strokes form the basis of all Vertical Gothic Lettering. The three basic strokes are: straight (vertical, horizontal, slant), left curve, and right curve. The basic principles of these strokes are demonstrated in figure 6-3. To differentiate still further, the basic strokes can be separated into nine subdivisions: vertical, horizontal, left slant, right slant, left curve, right curve, top curve, bottom curve, and "S", as shown in figure 6-3.

   c. *Direction of Brush Strokes.* The appearance of a hand-drawn letter depends, to a very considerable degree, upon the direction given to each brush stroke. It is therefore important to closely follow the standard directions shown in figure 6-4 using the oval wash brush.

   d. *Right and Wrong Ways of Lettering.* Avoid the mistakes indicated in figure 6-5, and follow the right methods shown.

   e. *Spacing and Balance.* It is particularly important for the less experienced sign painter to pencil-in the letters upon the working surface before painting, making sure that they are accurately spaced and balanced and of uniform size and relationship. It may be necessary to letter under difficult conditions, at times, and with limited materials. In this event, the following method should be used.

THICKNESS IS APPROXIMATELY
1/6 TO 1/7 THE HEIGHT
OF THE LETTER

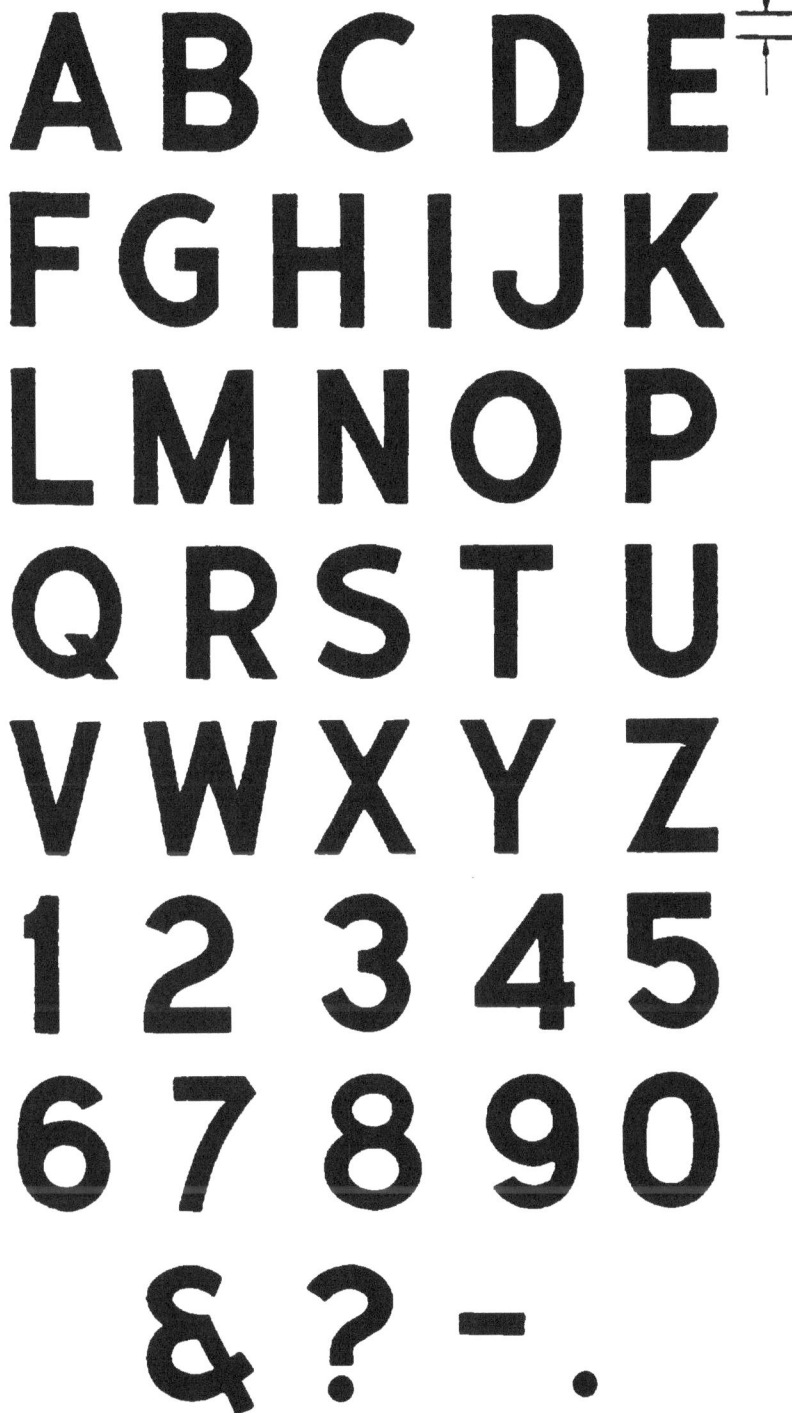

ABCDE
FGHIJK
LMNOP
QRSTU
VWXYZ
12345
67890
& ? - .

Figure 6-1. Lettering and Stencil Alphabet

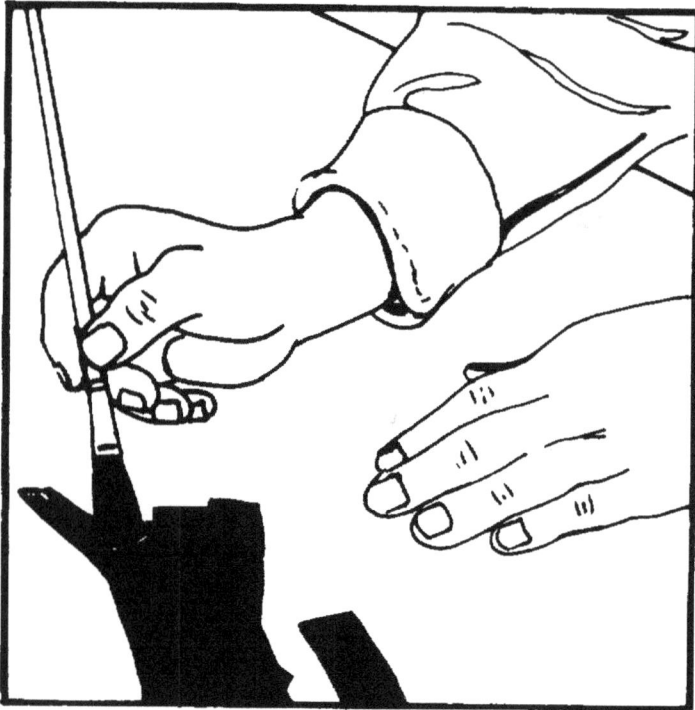

A.

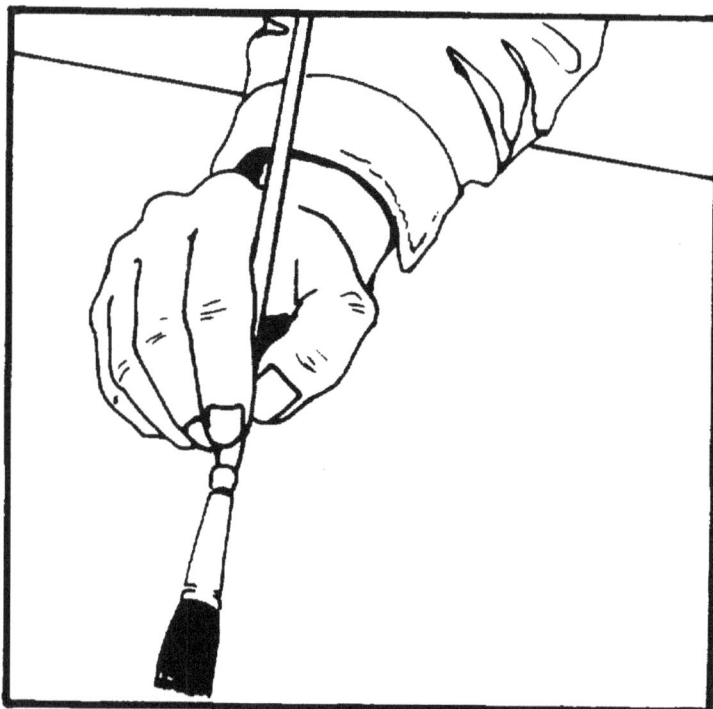

B.

Figure 6-2. Brush Preparation for Lettering

BRUSH
CENTER OF
BALANCE

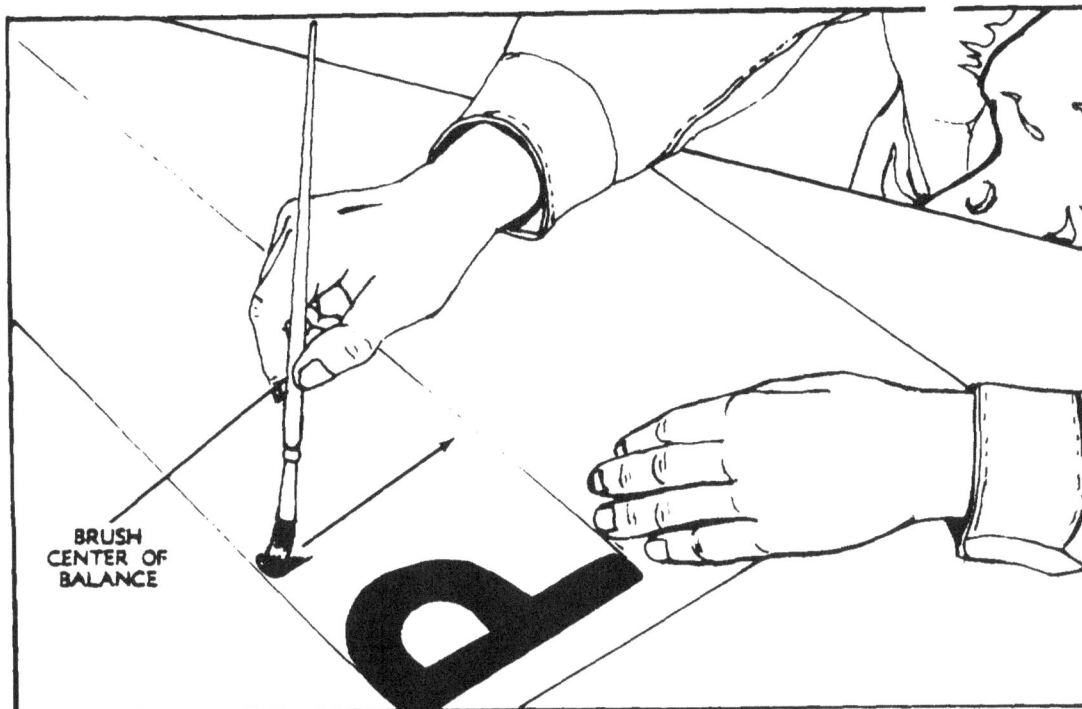

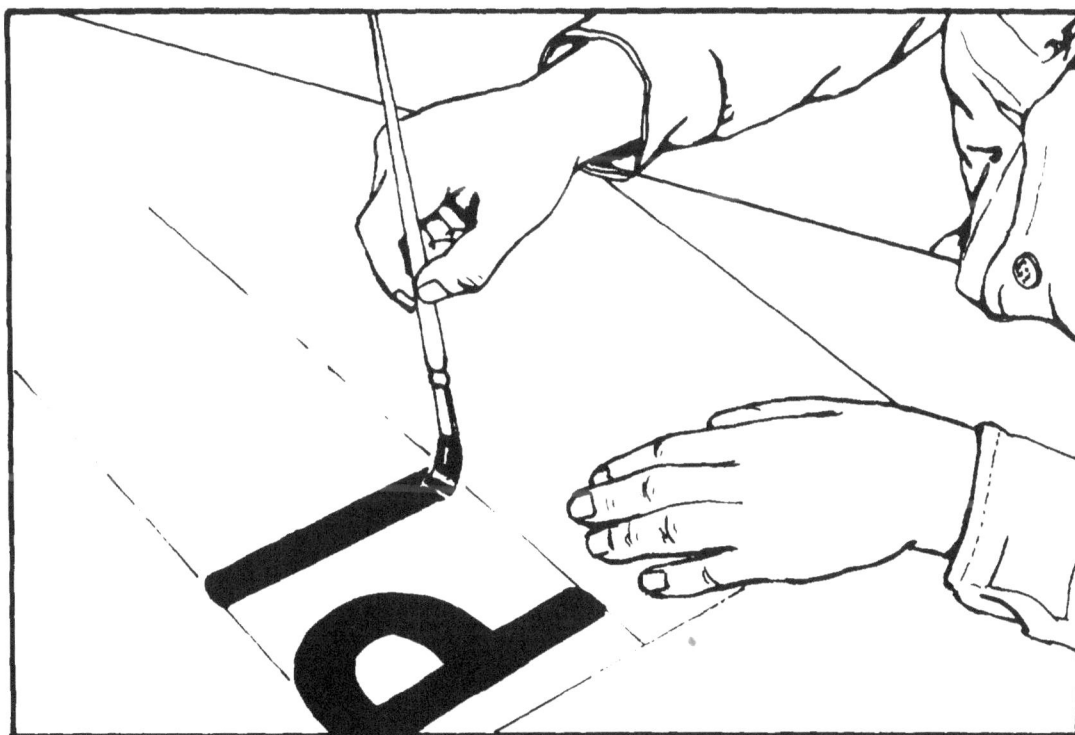

Figure 6-3. Lettering Techniques (Sheet 1 of 3)

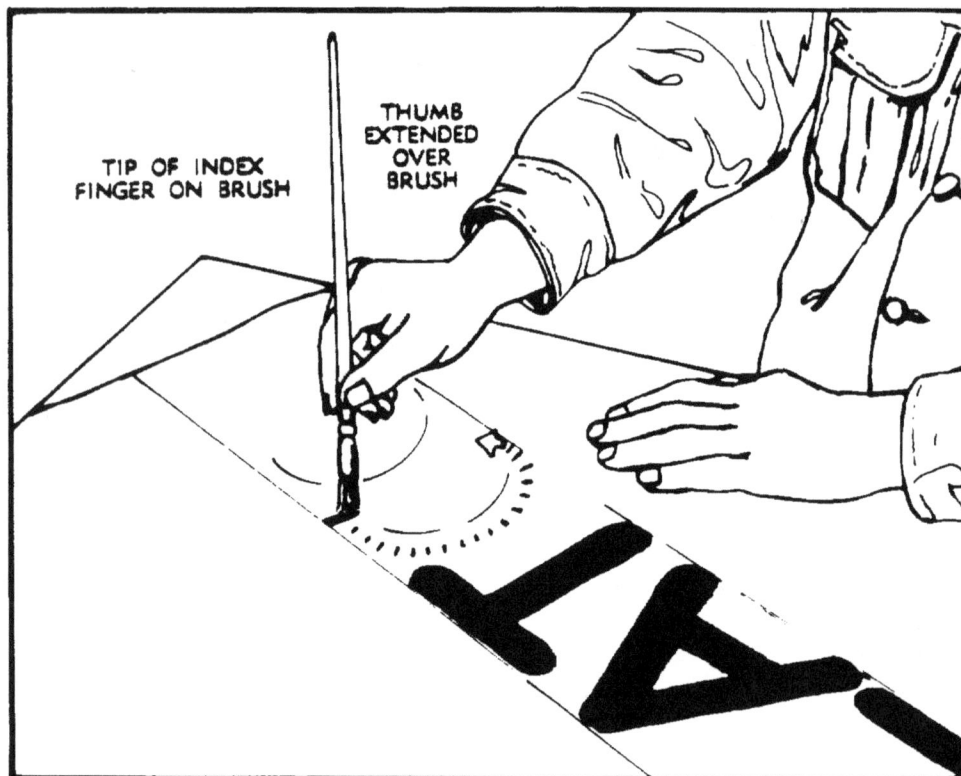

A. LETTERING—METHOD OF HOLDING BRUSH AT START OF LEFT CURVED STROKE.

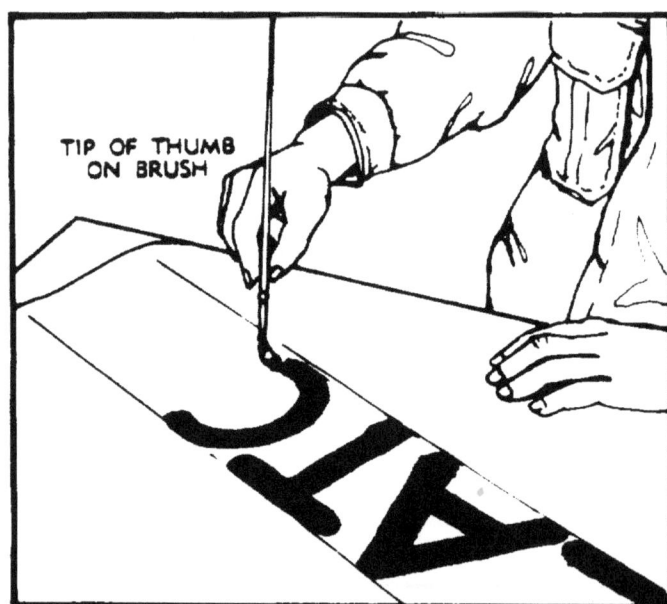

B. LETTERING—POSITION OF BRUSH AT END OF LEFT CURVED STROKE.

Figure 6-3. Lettering Techniques (Sheet 2 of 3)

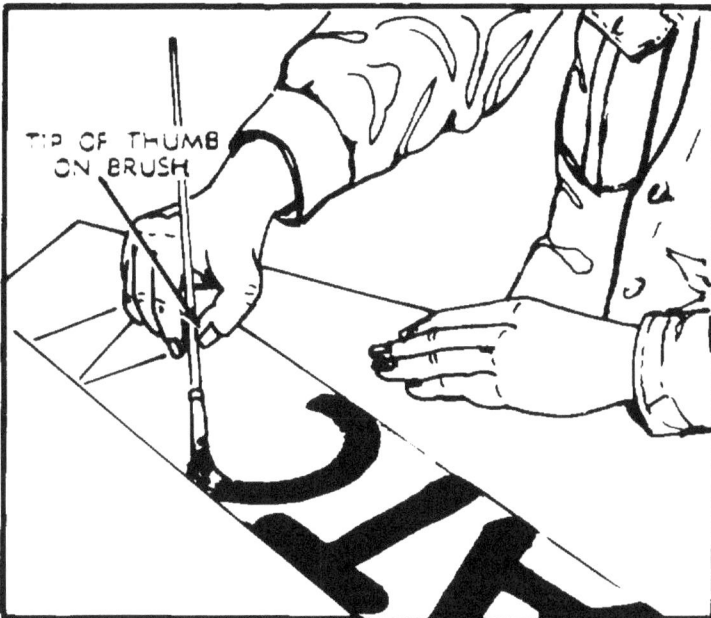

C. LETTERING—METHOD OF HOLDING BRUSH AT START OF RIGHT CURVED STROKE.

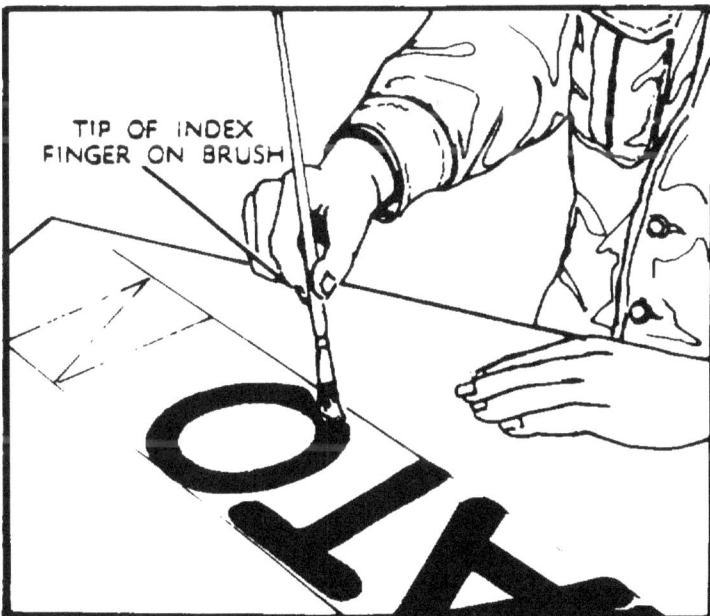

D. LETTERING—POSITION OF BRUSH AT END OF RIGHT CURVED STROKE.

Figure 6-3. Lettering Techniques (Sheet 3 of 3)

BASIC BRUSH STROKES

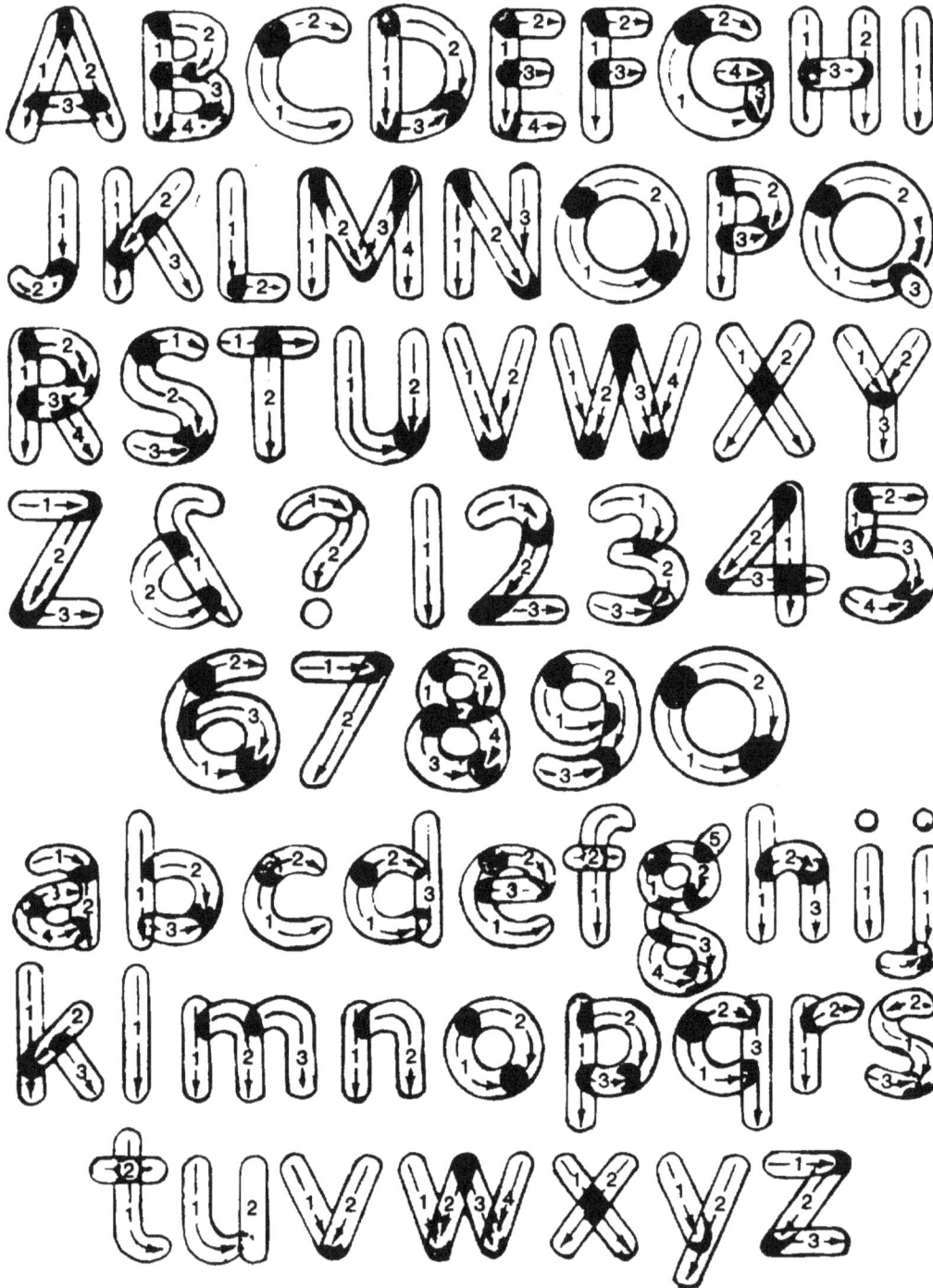

Figure 6-4. Brush Strokes for Lettering

6-7

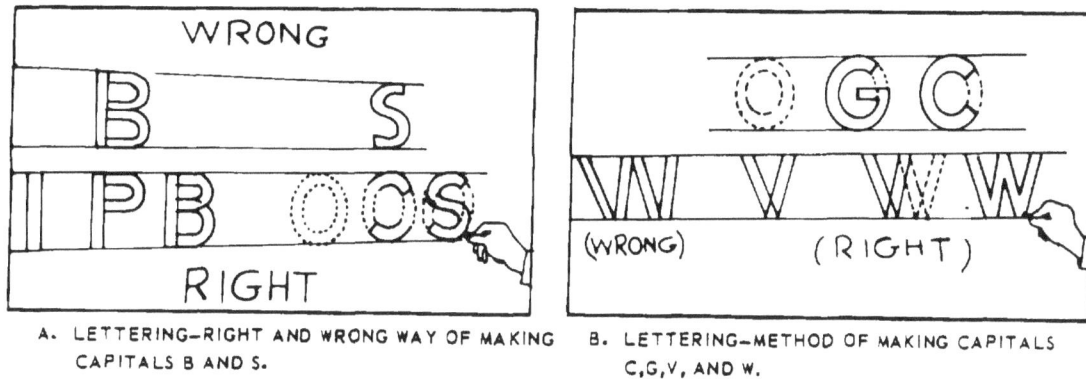

A. LETTERING—RIGHT AND WRONG WAY OF MAKING CAPITALS B AND S.

B. LETTERING—METHOD OF MAKING CAPITALS C,G,V, AND W.

Figure 6-5. Lettering Method

6-5. HAND SIGN PAINTING

a. Identify the area to be marked and clean it of dirt, grease, and base paint. Using chalk, draw parallel lines the width of the desired letters and numbers, as shown in figure 6-5.

b. Using the techniques described above, draw letters and numbers, with chalk, between parallel lines in preparation for painting.

c. Paint the letters and numbers using the proper paint brush and paint.

d. Allow the paint to dry before touching. The time for drying is dependent upon the kind of paint used, the temperature, and the thickness of the paint film. Protect the markings from dust and dirt until dry.

e. Once thoroughly dry, use a cloth and rub off the chalk guidelines.

Section II. STENCIL AND PAINT MAKING

6-6. PURPOSE

Stencils enable untrained personnel to apply lettering and designs to materiel quickly and efficiently. A stencil is a paper or metal pattern which has the letters or design cutout, so that when the stencil is held in position over a surface and paint is applied to the cutout portions, the lettering or design will be accurately reproduced. When a large number of signs, identification marks, or designs are to be reproduced, time is saved by using a stencil.

6-7. STENCILING TECHNIQUES

a. General.

(1) This method of painting requires the use of gummed-back (pressure-sensitive) paper stencils.

(2) Paper stencils are available as individual letters, numerals, and legends of various sizes.

(3) The surface to which the marking is to be applied must have all oil, dirt, and grease removed and must be dry to prevent contamination of the stencil adhesive and the marking paint. Use liquid detergent cleaner, MILD-16791, Type I, or another approved cleaning solution for this purpose.

6-8

b. Paper Stencil Application Techniques.

(1) Identify the area to be marked. Draw a straight guideline for proper positioning of the letters, numbers, or legend.

(2) Peel off gum-protector paper from the paper stencil; avoid touching the adhesive.

(3) Handle the stencil with caution to avoid wrinkling or distorting the characters, and apply to the proper location on the vehicle or equipment.

(4) After fixing the stencil in place, remove the webs from the letters and numerals so that the finished marking appears with unbroken lines.

(5) Mask the areas between, above, and below the stencils using masking tapes and paper to prevent overpainting.

(6) Apply paint to the stenciled area by spraying or brushing; spraying is preferred. Use CARC lusterless Black 37030 or 37038, or Green 383 (from table 4-1 in Chapter 4) for applications requiring chemical agent resistant systems.

(7) A few minutes after painting, remove the stencils. This must be done with care to avoid smudging the marking or the adjacent surface area.

(8) Do not touch the painted marking until thoroughly dry; drying time is dependent upon temperature, type of paint, method of application, and the thickness of the paint film. Take care to protect the marking from dust and dirt during the drying period.

(9) Carefully clean all paint overspray, smudges, and residue from the area. Use mineral spirits paint thinner and a cloth. This must be done with care; avoid contact of the marking with the thinner.

Section III. PRESSURE SENSITIVE ADHESIVE VINYL MARKERS (DECALS)

6-8. GENERAL

a. These markers are available as die-cut letters, numerals, and legends in various sizes.

b. The vinyl markers are applied to the surface without water or other solvent to activate the adhesive. They are mounted on a protective paper lining that is removable without the use of a solvent. The marker face is covered by a translucent application tape which is removed after marker application.

c. The markers are resistant to grease, oil, water, salt spray, gasoline, and aromatic fuels. Cleaning of the marking requires only water and soap or detergent.

6-9. APPLICATION TECHNIQUES

a. Surface. The surface on which the marker is to be applied must be clean and dry. All oil, grease, and dirt must be removed by washing with liquid detergent cleaner or an approved cleaning solution to prevent contamination of the adhesive. Allow sufficient time for the cleaning agent to evaporate. Vinyl markers cannot be adequately applied to extremely irregular or rough surfaces. Complete contact of the marker to the surface is necessary for proper adhesion. Old markers must be removed completely prior to applying a new marker.

(1) Old pressure sensitive vinyl markers may be removed by soaking the markers with a rag or sponge dipped in technical methyl ethyl ketone or technical xylene, TT-X-916, which acts on the adhesive to soften it. A mixture of 75 percent methyl ethyl ketone and 25 percent technical xylene is recommended. The marker may then be removed with a putty knife or scraper without damaging the materiel surface.

WARNING

Methyl ethyl ketone and technical xylene are hazardous substances and must be disposed of in accordance with AR 420-47, Solid and Hazardous Waste Management.

(2) Alternate but less effective methods of removing the markers involve the use of common paint removers or a sharp bladed instrument. Avoid materiel surface damage and self-inflicted personal injury.

CAUTION

The vinyl marker cannot satisfactorily be removed by power sanding or abrasion. These methods will damage the materiel surface, necessitating refinishing.

b. Temperature. Application of the vinyl marker should be made at moderate temperatures above 40°F (4.44°C), but may be made at lower temperatures if the surface is prewiped with technical isopropyl alcohol. If the surface temperature is warm or hot, insure that application is exact at first contact since the decal will stick quickly.

c. Sealing. Sealing of the marker or its edges with varnish or other sealant is neither required nor recommended.

d. Legend Marker Application.

(1) Each legend marker is prespaced and precentered on the application tape and backed with a protective liner over the pressure sensitive adhesive.

(2) Mark a straight horizontal guideline on the materiel surface in the appropriate location. This guideline will be used for properly positioning the legend.

(3) Place the legend on a flat surface with the translucent application tape side down and carefully remove the protective liner. Avoid handling the adhesive on the legend marker.

(4) Position the legend to the guideline on the materiel. Press one edge down while holding the rest of the legend taut and slightly away from the surface (see figure 6-6A).

(5) Roll the legend down firmly with a roller or applicator to remove any trapped air bubbles or wrinkles (see figure 6-6B).

(6) Starting at one corner of the marker, remove the application tape by carefully peeling it back flat against itself (see figure 6-6C).

(7) Roll the legend again to insure complete and firm adhesion.

e. Character (Letter or Numeral) Marker Application.

(1) Each character (letter or numeral) is precentered on the application tape and backed with a protective liner over the pressure sensitive adhesive.

(2) Mark a straight horizontal guideline on the materiel surface in the designated location. This guideline will be used for proper alignment of the characters.

(3) Place the first character on a flat surface with the translucent application tape side down and carefully remove the protective liner. Avoid handling the adhesive on the character.

(4) Position the character to the guideline on the materiel. Press one edge down while holding the rest of the character taut and slightly away from the surface.

(5) Roll the character down firmly with a roller or applicator to remove any trapped air bubbles or wrinkles. Do not remove application tape at this time.

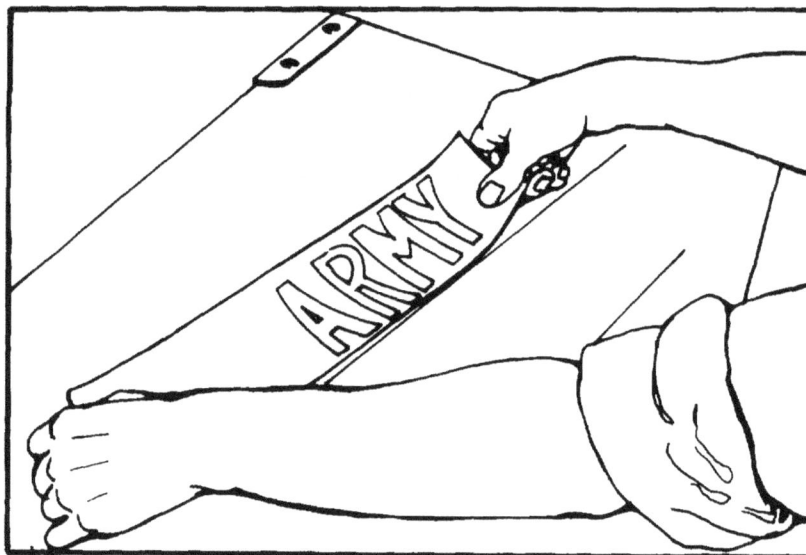

A. ALINE LEGEND TO GUIDELINE.

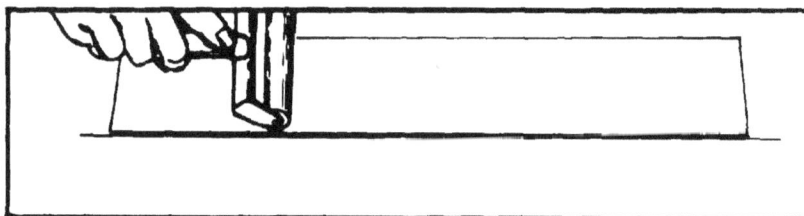

B. REMOVE AIR BUBBLES AND WRINKLES.

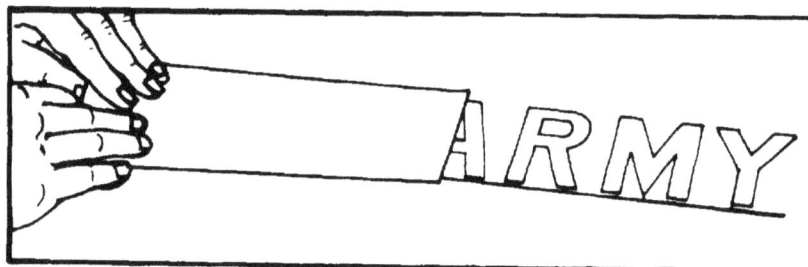

C. REMOVE APPLICATION TAPE.

Figure 6-6. Applying Adhesive Vinyl Markers (Decals)

(6) Repeat steps (3), (4), and (5) above, in order, for each remaining character in the desired marking. Place the left edge of the application tape against the right edge of the preceding application tape (see figure 6-7A).

(7) When the entire marking is properly positioned and applied, remove the application tape. Start at a corner and carefully peel each application tape back, flat against itself (see figure 6-7B).

(8) Roll the characters again to insure firm adhesion.

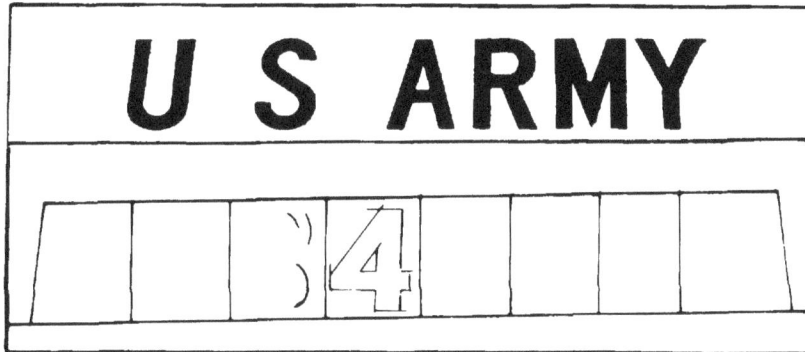

A. CHARACTERS SPACED PROPERLY.

B. REMOVE EACH TAPE SEPARATELY.

Figure 6-7. Letter or Numeral Application

6-10. VEHICLE NATIONAL SYMBOL MARKINGS

a. This paragraph is concerned with the application of the National Symbol (star) to vehicle surfaces. This method applies to National Symbols made of vinyl material.

b. The National Symbol markings are available in various sizes from six to 36 inches (measurement between opposite points).

c. The vinyl material National Symbol is applied directly to the equipment surface without the use of water or other solvent to activate the pressure sensitive adhesive. The symbol is mounted on a protective liner with the symbol face covered by a premask tape.

d. The vehicle or equipment surface must be cleaned of dirt, grease, dust, and loose paint prior to application.

e. Application.

(1) Place the symbol on a flat surface, face up. Cover one point of the symbol with a small piece of masking tape, rubbing the tape down firmly onto the symbol.

(2) Hold the symbol by the tape, in one hand. Begin separation of the protective paper liner from the adhesive side of the symbol.

(3) Place the symbol on a flat surface, face down. Carefully continue pulling the paper liner from one point of the symbol past the horizontal base of the point. Fold the liner as it is freed from the symbol.

(4) Position the symbol on the equipment surface. Apply the exposed symbol tip to the surface while holding the rest of the symbol taut and slightly away from the surface. Apply the exposed portion of the symbol while rolling and pressing the material to remove wrinkles and air bubbles.

(5) Continue removing the paper liner as stated in (3) and (4) above, rolling and pressing the unapplied portion of the symbol to the surface until the entire marking is applied.

(6) Roll the entire marking again, with particular attention to the edges, to insure firm and complete adhesion.

(7) Remove the premask tape on the face of the symbol by pulling carefully on the masking tape piece (applied in (1) above), folding the premask tape back against itself. Carefully pull back to the opposite edge of the symbol. With this operation, the protective premask tape will tear. The remaining pieces may be removed by pulling them, folded back, from the center of the symbol to the remaining symbol points. Roll the marking again with particular attention to the edges.

(8) Any remaining small air bubbles may be punctured with a pin and the air may be worked out with a finger.

Section IV. POUNCING

6-11. DESCRIPTION AND PURPOSE

a. When it is necessary to make a quantity of the same legends, signs, identification marks, or designs, and a stencil legend is not already available, work can be speeded up by a process known as pouncing. Pouncing is the term applied to the use of a perforated pattern in transferring the outline of the legend, sign, or design to be painted to the painting surface.

b. Pouncing should also be used when more accurate lettering and designs are desired than can be attained by stenciling letters individually, and particularly when working over larger areas.

6-12. Equipment

The following equipment and materials are needed to prepare a pouncing pattern:

a. Thin, durable paper (large enough to cover the lettering or design).

b. Light cardboard.

c. A pouncing wheel.

d. Dry color, powdered chalk, or other powder.

e. Flint sandpaper, grade 2/0.

f. Masking tape.

g. A thin cloth.

6-13. PROCEDURE

a. Pencil-in (draw) the letters, numerals, or design on a plain sheet of paper.

b. Place the penciled-in paper on top of cardboard or other material which can be easily perforated by a pouncing wheel. Then, by using a pouncing wheel, perforate the outline of the markings which have been penciled-in (See figure 6-8A). If a pouncing wheel is not available, use a large needle or other sharp pointed object to perforate the outline.

c. Turn the pattern over and use flint sandpaper, grade 2/0, to sand off all rough edges of the perforations (see figure 6-8B).

d. Prepare a pouncing bag by placing dry color, powdered chalk, or any other available powder in a thin cloth. Tie the cloth so it forms a bag with the powder inside.

e. Place the pattern in the desired position on the surface to be painted. Secure it in position with masking tape. Then gently tap all perforations with the pouncing bag until the powder is worked through the perforations and onto the painting surface (see figure 6-8C).

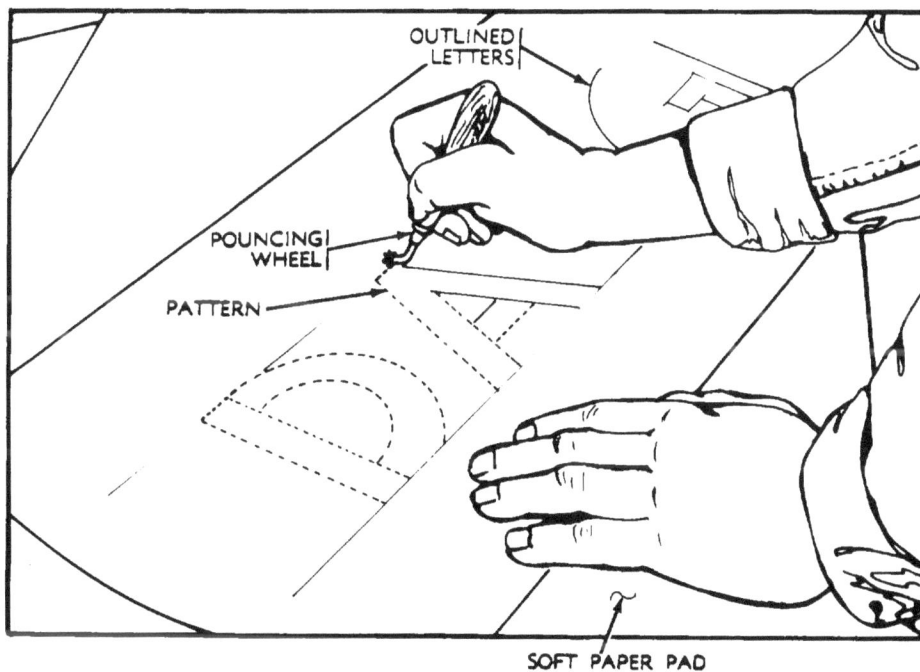

A. PERFORATING A PATTERN FOR POUNCING

Figure 6-8. Pouncing Techniques (Sheet 1 of 3)

B. SANDING THE BACK OF A POUNCING PATTERN.

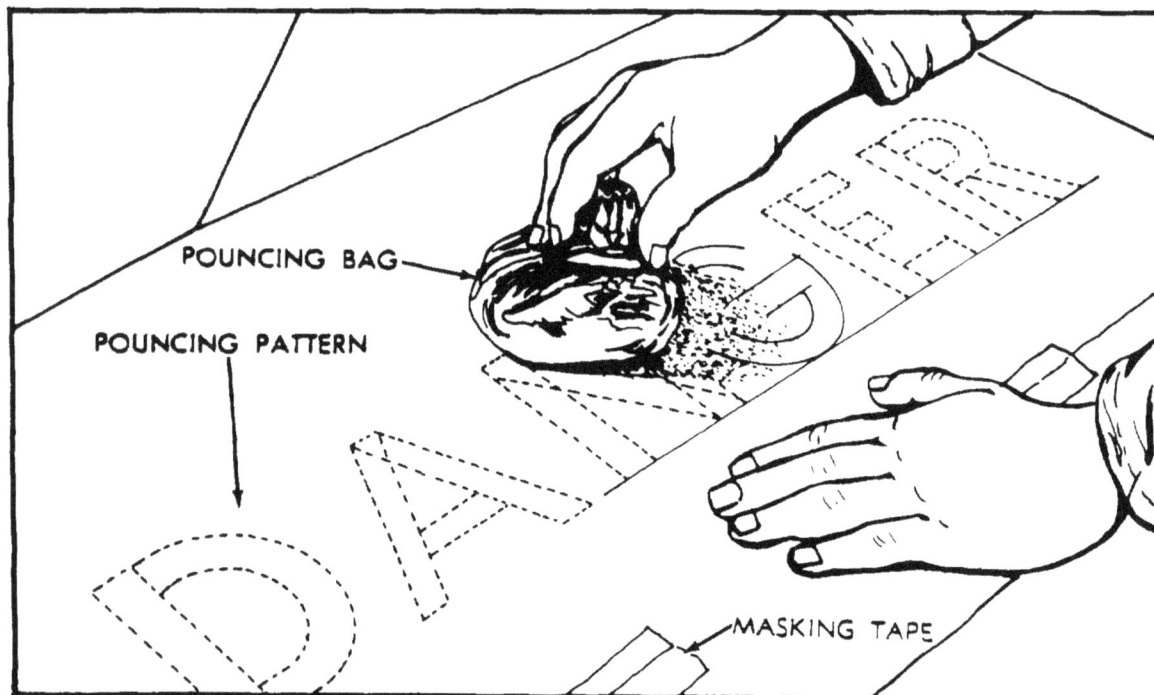

C. POUNCING A LETTERING PATTERN.

Figure 6-8. Pouncing Techniques (Sheet 2 of 3)

f. Remove the pattern (see figure 6-8D), taking care not to smudge the perforation dots on the painting surface.

g. Select the proper paint brush and paint in the legend or design (see figure 6-8E), taking care not to go outside the dotted pattern.

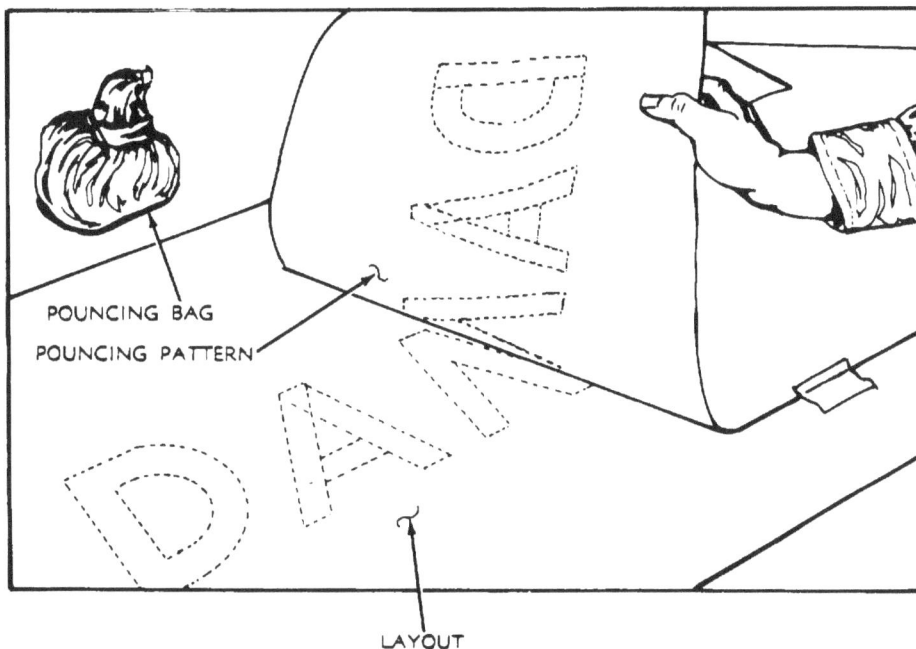

D. REMOVING THE PATTERN. POUNCING PATTERN TRANSFERRED TO SURFACE TO BE PAINTED.

E. PAINTING IN A POUNCED SIGN PATTERN

Figure 6-8. Pouncing Techniques (Sheet 3 of 3)

h. Allow the paint sufficient time to dry before touching it. Drying time is dependent upon the kind of paint used, the temperature, and the thickness of the paint film. Protect the markings from dust and dirt until dry.

i. Once the paint is thoroughly dry, use a cloth and rub off any leftover powder residue.

Section V. SILK SCREEN PRINTING PROCESS

6-14. GENERAL

a. The silk screen process is a method of printing, in one or more colors, on almost any type of flat surface with components such as oil colors, water colors, lacquers, enamels, and polyurethanes. This process is used when large quantities of the same design are desired in a relatively short time.

b. The basic principles of this process are simple but require special equipment. Care and accuracy in performing preparatory work will assure the success of the operation.

6-15. EQUIPMENT REQUIRED

a. Baseboard. A baseboard must be larger than the design, with a surface that is absolutely level and flat. This board may be a drawing board, ply board, or table-top, and its surface should be covered with stiff cardboard. If either the baseboard or its mating screen frame become warped, it will be impossible to produce an acceptable print.

b. Frame. The frame on which the silk, or a synthetic such as nylon, is attached may be a simple wooden frame or the standard grooved frames stocked by artist supply dealers in various sizes. This frame should be at least one and one-half times longer and wider than the size of the image that is to be printed (see figure 6-9).

c. Silk. Silk or synthetic-silk type material should be used for screening. This material is available in different meshes. For best results, use the "medium size" mesh or numbers "14XX" or "16XX". The double "X" denotes that the material has a double weight or double strength rating.

d. Loose Pin Hinges. The frame should have a pair of loose pin hinges attached to one of the long sides of the frame and to the baseboard. The hinges are placed so that the silk surface of the frame lies in flat contact with the baseboard. Masking tape is then placed on the inside of the frame so that half of the tape's width is on the frame and the other half is on the silk. The masking tape will prevent the paint from oozing between the silkscreen and frame and onto the print and/or baseboard.

e. Tacking. The material must be stretched drum-tight over the frame and secured by tacking to the frame's outer surface. The material or silk is then washed with water to remove the sizing and facilitate tightening. Failure to achieve the drum-tightness will result in unsatisfactory reproductions.

f. Squeegee. The squeegee consists of a rubber straight edge embedded in a length of wood that has clearance with one of the interior dimensions of the frame, and is shaped to conveniently fit the hands. The squeegee must be at least two inches larger than the width of the design being reproduced so that one pull over the screen will complete the printing. To assure flat contact with the silk, draw the rubber edge of the squeegee across a piece of flat sandpaper while holding the squeegee in a vertical position.

g. Other Materials. Stencil knives, dividers, a steel rule, a T-square, a triangle, translucent paper-backed film, and adhesive tape are also necessary items for the silk screening process.

6-17

HINGE · SILK SCREEN · FRAME

PRINT

SPLINE

A. SILK SCREEN, PRINT ARRANGEMENT

RETAINING BEAD

SQUEEGEE

INK

HINGE

FRAME

SILK SCREEN

TAPE

DANGER

PENCIL
LINE

PRINT

DIRECTION OF
SQUEEGEE
PULL

REGISTRATION
MARK

TACK

B. PRODUCING A "DANGER" PRINT

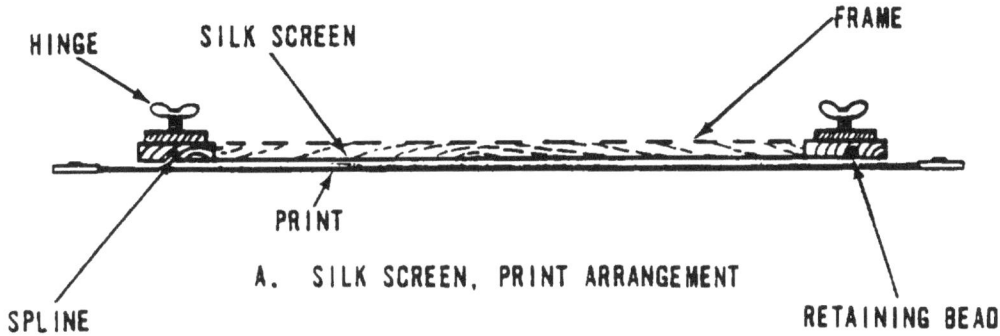

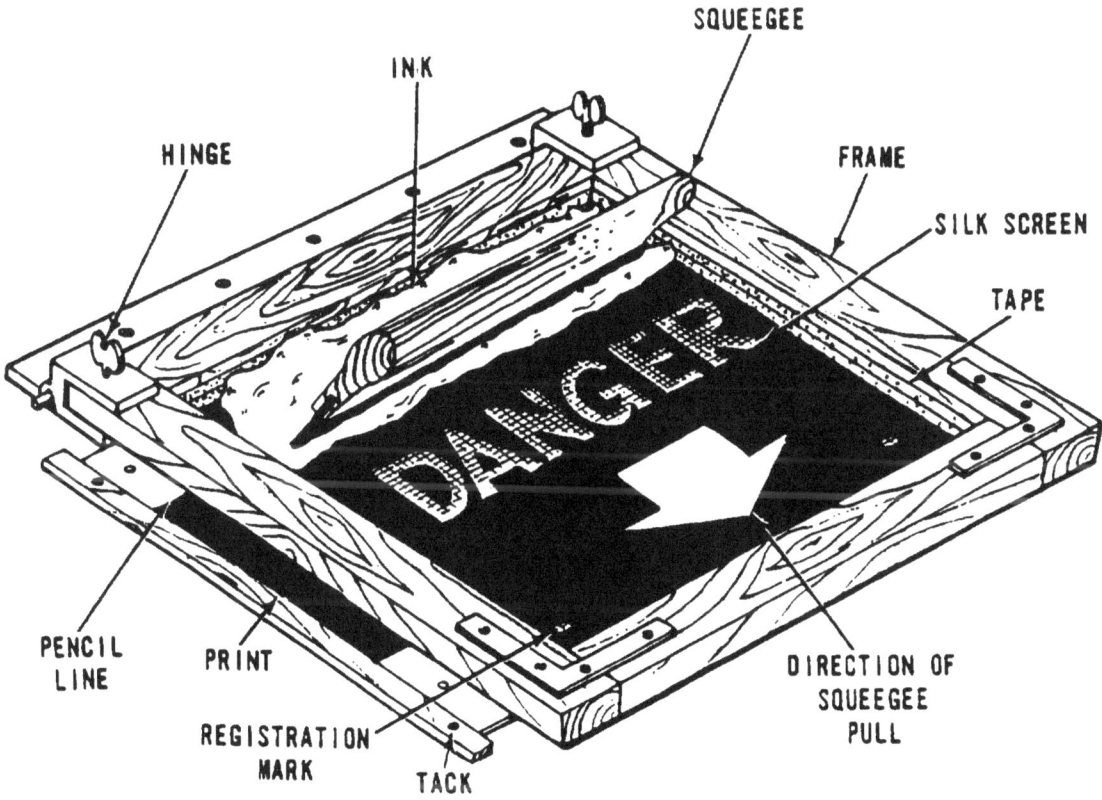

Figure 6-9. Silk Screen Process

6-16. SPECIFIC INSTRUCTIONS

a. *Preparatory Work.*

(1) Thumbtack the original sketch or layout sheet to either a drawing board or worktable. A drawing board will be found to have many advantages over a worktable as the operator can turn it at will and sit in a more comfortable position during the cutting operation.

(2) With pencil or pen draw a "cross" in each of the four corners of the original sketch to help in replacing the film in the exact location each time cutting is interrupted.

(3) Cut a piece of film for each of the colors found in the sketch, making each a little larger than the sketch.

(4) With adhesive tape, fasten the film over the original sketch with the film side up and the translucent paper side down, in contact with the sketch. Use enough tape so that the film is firmly held in place and will not shift. Using a pen, trace the "cross" appearing in each of the four corners of the sketch onto the film. The film is now ready for cutting.

b. *Film Cutting.*

(1) This specially prepared film is laminated to a sheet of translucent paper (the paper acts only as a temporary carrier of the film until such time as the transfer to the silk is made). The proper method is to cut only through the film and not through the backing paper.

(2) The order in which the colors are to be processed must be determined before any cutting occurs and may not thereafter be changed. The general practice is to process the lightest color first and the darkest color last. The makeup of the sketch occasionally requires a change from this procedure.

(3) The cutting operation must be performed in a manner to allow the first color applied to extend under the edge of the succeeding colors. In a tracing manner, using a sharp stencil knife, cut through the film to the backing paper, but not through the backing paper (an hour or so of practice in cutting should be sufficient to become used to this procedure). When the cutting has been completed, strip out the film representing the portion of the design to be applied in the first color. When finished, remove this cut film sheet from the sketch and place carefully to one side. Repeat this operation for each remaining color. Particular attention must be given to the accuracy of the cutting which follows along the lines in the original sketch, as these must interface and mate with the masks of the other colors in the reproduction. To prevent blurring of corners the cuts should extend just past the intersections.

c. *Attaching the Film to the Screen.*

(1) In all the larger open spaces from which the film has been stripped, cut a slit through the backing paper. The purpose of this is to allow air to escape during the adhering operation.

(2) Place the silk screen frame in the hinges on the printing table. Secure the layout sheet to the table and apply registration strips of the same thickness as the material that is to be imprinted. These strips should fit closely with the aforementioned material. Bring down the screen so that the silk is in contact with the imprintable material and make certain that the contact is good. If the contact is poor, build up the surface by using a piece of paper, or other material, under the layout sheet. Replace the film for the first color, taking care to align the registration marks with those on the layout sheet, and fasten the layout sheet with adhesive tape.

(3) Examine the stencil to make sure that no small pieces of film have been left in the cut portions.

(4) Obtain two pieces of soft cotton cloth, one large and one small. Do not use cheesecloth or similar cloths as difficulty will be encountered in judging the amount of liquid applied. Roll up the large pieces in a loose ball and wet the small piece with the adhering liquid. With the silkscreen frame resting on the first color film, dampen (do not soak) a small portion of the screen by taking a single stroke and dry it immediately with the dry

cloth using a rubbing motion. When this has been done, adhesion will be instantaneous. Continue in the same manner until the entire film has been adhered, wetting the small cloth as often as it is necessary. In adhering, always start from one side of the screen and continue in the same direction to avoid wrinkles.

(5) When the entire film has been adhered, take a thin straightedge or ruler and slip it under the film to carefully free the screen from the layout sheet while loosening the restraining adhesive tape. Remove the silk screen frame from the hinges and lay it on the table with the film paper side up. Allow the film to dry about 10 minutes.

d. *Removing the Backing Paper.*

(1) Start in any one of the four corners and slowly peel off the backing paper. Peel the backing paper so that one can, at all times, see the film in order to prevent tearing any portion of it that has not adhered. Should any portion of the film not adhere properly, do not remove the backing paper, but turn the screen over again and wet and dry that portion (as in paragraph c.(4) above) to obtain the proper adhesion.

(2) When the backing paper has been completely removed, if there are still some loose places, wet the cloth with the adhering liquid and dampen the loose part from the under side of the frame. Pat the loose portion down from the film side, thus completing the adhesion.

(3) Fill in the open silk bordering the film with lacquer.

e. *Printing or Reproducing.*

(1) If the original sketch is the exact size of the copy to be processed, the original registration strips applied should suffice. If this is not the case, new registration strips should be set in place without disturbing or moving the layout sheet.

(2) Place one of the pieces to be printed on the table, snugly against the registration strips, and lower the silk screen.

(3) Place a small quantity of the first color of paint across the screen just above the design. Starting just above the paint, pull the squeegee across the screen while making certain to apply firm and even pressure across the ,width of the squeegee (see figure 6-9).

(4) Lift the screen, remove the printed piece, and place it on a drying rack, as illustrated in figure 6-10. To continue the process, insert an unpainted piece, lower the screen, and squeegee in the opposite direction.

(5) When the first color has been applied to all of the pieces that are to be printed, the film or mast must be removed from the silk and replaced with the film that was cut for the second color.

f. *Removing Film from the Silk.*

WARNING

Avoid skin contact with lacquer thinner. It can cause a skin rash.

WARNING

Lacquer thinner is highly flammable. Keep away from flames and sparks.

6-20

Lay a sheet of smooth wrapping paper on a flat table, placing the silk screen frame on top with the film side down. Soak a cloth with lacquer thinner and wipe over the film portions of the screen while keeping the rag well saturated. After the lacquer has softened, pull the paper and the attached film away from the silk screen. Use several clean cloths to alternately wash the old lacquer from the silk screen. Dry the silk with a soft cloth.

    g. Facts Pertinent to the Silk Screen Printing Process.

       (1) When a job is to be repeated at frequent intervals, the silkscreen frames, with cut flm applied, may be stored for future use.

       (2) In the handling and storage of silk screen frames, care must be exercised to avoid stretching or puncturing the silk.

       (3) , an average of 5,000 to 10,000 impressions may be made before the silk wears out.

       (4) Experience will indicate the proper consistency at which the paint should be applied. Because solvent continually evaporates, it is customary to have the replenishing paint slightly thinner than the first portion put into the screen. This procedure brings the consistency of the paint back to that desired immediately after replenishment.

       (5) There are various types of space-saving drying racks employed in silk screening. Figure 6-10 illustrates a typical unit.

Figure 6-10. Drying Rack

6-21/(6-22 blank)

APPENDIX A
REFERENCES

AMC-P-750-9	Maintenance of Supplies and Equipment
AR 40-5	Preventive Medicine
AR 200-1	Environmental Protection and Enhancement
AR 200-2	Environmental Effects of Army Actions
AR 381-143	(C) Logistic Policies and Procedures, (U)
AR 385-30	Safety Color Code Markings and Signs
AR 385-63	Policies and procedures for Firing Ammunition for Training, Target Practice and Combat
AR 420-47	Solid and Hazardous Waste Management
AR 708-1	Cataloging and Supply Management Data
AR 750-1	Maintenance of Supplies and Equipment, Army Materiel Maintenance Policies
AR 750-4	Maintenance of Supplies and Equipment-Depot Materiel Maintenance and Support/Training Activities
AR 840-10	Flags, Guides, Streamers, Tabards, and Automobile and Aircraft Plates
ASTM D846	Xylene, Standard Specification for Ten Degree
CFR 1910	OSHA Safety and Health Standards
DOD-P-15328D	Primer (Wash), Pretreatment (Formula No. 117 for Metals) (Metric)
DODI 4145.19-R.1	Hazardous Materials Handling and Storage Criteria
DOT AC 70/7460-1	Obstruction, Marking and Lighting
FAA TSO C26	Aircraft Wheels and Brakes
	Federal Specification L-S-300
	Sheeting and Tape, Reflective: Nonexposed Lens, Adhesive Backing
Fed Std 595	Colors
FM 5-20	Camouflage
FM 55-30	Army Motor Transport Units and Operations
MIL-A-8625	Anodic Coatings, for Aluminum and Aluminum Alloys
MIL-C-450	Coating-Compound, Bituminous Solvent Type, Black (for Ammunition)
MIL-C-5541	Chemical Conversion Coatings on Aluminum Alloys
MIL-C-8514	Coating Compound, Metal Pretreatment, Resin-Acid
MIL-C-10578	Corrosion Removing and Metal Conditioning Compound (Phosphoric Acid Base)
MIL-C-22750	Coating, Epoxy Polyamide
MIL-C-46168	Coating, Aliphatic Polyurethane, Chemical Agent Resistant
MIL-C-53039	Coating, Aliphatic Polyurethane, Single Component, Chemical Agent Resistant
MIL-C-53072	Chemical Agent Resistant Coating (CARC) System Application Procedures and Quality Control Inspection
MIL-C-85570	Cleaning Compound, Aircraft, Exterior
MIL-D-16791	Detergent, General Purpose (Liquid, Nonionic)
MIl-D-23003	Deck Covering Compound, Nonslip, Lightweight
MIL-E-7125	Ethylene Glycol Monoethyl Ether Acetate, Technical
MIL-F-14072	Finishes for Ground Electronic Equipment
MIL-L-11195	Lacquer, Lusterless, Hot Spray
MIL-L-12277	Lacquer, Automotive, Hot Spray
MIL-L-46159	Lacquer, Acrylic, Low Reflective, Olive Drab
MIL-L-52043	Lacquer, Semigloss, Cellulose Nitrate

MIL-M-3171	Magnesium Alloy, Processes for Pretreatment and Prevention of Corrosion On
MIL-M-13231 (ER)	Marking of Electronic Items
MIL-M-43719	Marking Materials and Markers, Adhesive, Elastomeric, Pigmented Legends
MIL-N-15178	Solvent, Naphtha
MIL-P-11414	Primer Coating, Lacquer, Rust Inhibiting
MIL-P-14105	Paint, Heat Resisting (for Steel Surfaces)
MIL-P-14458	Paint, Rubber, Red Fuming Nitric Acid Resistant
MIL-P-14631	Plate, Automobile, Distinguishing, General Office
MIL-P-15931	Paint, Antifouling, Vinyl (Formulas No. 121 and No. 129)
MIL-P-22636	Primer Coating, for Red Fuming Nitric Acid Resistant Paint
MIL-P-23377	Primer Coating, Epoxy Polyamide, Chemical and Solvent Resistant
MIL-P-24411	Paint, Epoxy Polyamide, General Specification for
MIL-P-52905	Paint, Arctic Camouflage, Removable
MIL-P-53022	Primer, Epoxy Coating, Corrosion Inhibiting, Lead and Chromate Free
MIL-P-53030	Primer Coating, Epoxy, Water Reducible, Lead and Chromate Free
MIL-P-53032	Primer Coating, Water Reducible, Epoxy Ester-Latex Type, Lead and Chromate Free
MIL-P-53044	Painting and Marking Freight and Maintenance Cars, Railway Motive Power and Work Equipment
MIL-P-85582	Primer Coating: Epoxy, VOC Compliant, Chemical and Solvent Resistant
MIL-R-81294	Remover, Paint, Epoxy, Polysulfide and Polyurethane Systems
MIL-S-5002	Surface Treatments and Inorganic Coatings for Metal Surfaces of Weapons Systems
MIL-S-11030	Sealing Compound, Non-Curing, Polysulfide Base
MIL-S-11031	Sealing Compound, Adhesive, Curing, Polysulfide Base
MIL-STD-171	Finishing of Metal and Wood Surfaces
MIL-STD-186	Protective Finishing for Army Missile Weapon Systems
MIL-STD-193	Paint Procedures and Marking for Vehicles, Construction Equipment, and Material Handling Equipment
MIL-STD-194	System for Painting and Finishing Fire-Control Material
MIL-STD-642	Identification Marking of Combat and Tactical Transport Vehicles
MIL-STD-709	Ammunition Color Coding
MIL-STD-1201	Alcohol, Denatured and Ethyl, Technical
MIL-STD-1247	Marking Functions and Hazard Designation of Hose, Pipe and Tube Lines for Aircraft, Missiles and Space Systems
MIL-STD-1473	Standard General Requirements for Color and Marking of Army Materiel
MIL-T-704	Treatment and Painting of Materiel
MIL-T-6095	Thinner, Cellulose-Nitrate-Dope, Blush Retarding
MIL-T-19544	Thinner, Aircraft Lacquer
MIL-T-81772	Thinner, Aircraft Coating
MIL-V-173	Varnish, Moisture and Fungus Resistant for Treatment of Communications, Electronic, and Assorted Equipment
MIL-W-5044	Walkway Compound, Nonslip and Walkway Matting, Nonslip
MIL-W-13518	Wood Preservative, Tetrachlorophenol and Pentachlorophenol, Surface Sealing Compound
NFPA Article 70	The National Electric Code
O-E-760	Ethyl Alcohol (Ethanol), Danatured Alcohol, Proprietary Solvents, and Special Industrial Solvents

Quadripartite Standardization Agreement 248	Identification of Medical Materiel to Meet Urgent Needs
SSPC-SP5-85	Steel Structures Printing Council Manual, Volume 2, White Metal Blast Cleaning
SSPC-SP6-85	Steel Structures Printing Council Manual, Volume 2, Commercial Blast Cleaning
SSPC-SP10-85	Steel Structures Printing Council Manual, Volume 2, Near-White Blast Cleaning
TB MED 502	Occupational and Environmental Health Respiratory Protection Program
TB MED 514	Occupational and Environmental Guidance for Painting Operations
TB 43-0118	Field Instruction for Painting and Preserving Electronics Command Equipment
TB 43-0144	Painting of Vessels
TB 43-0147	Color, Marking and Camouflage Patterns Used on Military Equipment
TB 43-0166	Color, Marking, and Camouflage Pattern Painting of Improved HAWK Guided Missile System Ground Support System
TB 43-0209	Color and Marking of Military Vehicles, Construction Equipment and Materials Handling Equipment
TB 43-0213	Corrosion Prevention and Control
TB 746-95-1	Color, Marking, and Camouflage Pattern Painting for Armament Command Equipment
TB 750-10	Painting, Replating and Preserving Instructions for Communications Security Equipment
TG 141	Industrial Hygiene Sampling Instructions
TM 5-200	Camouflage Materials
TM 9-1425-601-14	Color, Marking, and Camouflage Pattern Painting of PATRIOT Air Defense Guided Missile System Ground Support Equipment
TM 9-1425-2585-14	M48A2: General Maintenance Manual for M48A2: General Maintenance; Service Upon Receipt: Shipping and Storage and Demolition to Prevent Enemy Use, CHAPARRAL Air Defense Guided Missile System
TM 9-1430-588-20-1	Radar Set AN/MPQ-49, Forward Area Alerting Radar System
TM 55-1500-204-25/1	General Aircraft Maintenance Manual
TM 55-1500-345-23	Painting and Marking of Army Aircraft
TT-C-490	Cleaning Method and Pretreatment of Ferrous Surfaces for Organic Coatings
TT-C-520	Coating Compound, Bituminous, Solvent Type, Underbody (for Motor Vehicles)
TT-E-485	Enamel, Semigloss, Rust Inhibiting
TT-E-489	Enamel, Alkyd, Gloss (for Exterior and Interior Surfaces)
TT-E-522	Enamel, Phenolic, Outside
TT-E-527	Enamel, Alkyd, Lusterless
TT-E-529	Enamel, Alkyd, Semigloss
TT-L-215	Linseed Oil, Raw (for Use in Organic Coatings)
TT-P-28	Paint, Aluminum Heat Resisting (1200 Degrees F.)
TT-P-98	Paint, Stencil, Flat
TT-P-636	Primer Coating, Alkyd, Wood and Ferrous Metal
TT-P-645	Primer, Paint, Zinc Chromate, Alkyd Type
TT-P-664	Primer, Coating, Synthetic, Rust Inhibiting, Lacquer Resisting
TT-P-1757	Primer Coating, Zinc Chromate, Low Moisture Sensitivity

TT-R-251	Remover, Paint and Varnish
TT-S-300	Shellac, Cut
TT-S-720	Stain (Wood, Nongrain Raising, Solvent-Dye Type)
TT-T-291	Thinner, Paint, Mineral Spirits, Regular and Odorless
TT-T-306	Thinner, Synthetic Resin, Enamels
TT-T-548	Toluene, Technical
TT-T-801	Turpentine, Gum Spirits, Steam Distilled, Sulphate Wood, and Destructively Distilled
TT-V-51	Varnish, Asphalt
TT-V-121	Varnish, Spar, Water Resisting
TT-W-571	Wood Preservation, Treating Practices
TT-X-916	Xylene, Technical
US/GE Standard	Camouflage Pattern Painting
USAEHA Water Quality Information Paper No. 13	Pretreatment Regulations

**APPENDIX B
NSN TABLES**

B-1. SCOPE

This appendix contains tables of NSN's for various colors and sizes of coatings. It does not cover all coatings; however, most Chemical Agent Resistant Coatings (CARC) are covered. Also covered are some primers, Primer (Wash) Pretreatment DOD-P-15328, and Thinner MIL-T-81772. Table shows what each table covers.

Table B-1. NSN Tables

Table Number	Title
B-2	Coating, Aliphatic Polyurethane, CARC, MIL-C-46168, Type II
B-3	Coating, Aliphatic Polyurethane, CARC, MIL-C-46168, Type IV
B-4	Coatings, Aliphatic Ployurethane, CARC, MIL-C53039
B-5	Epoxy Coating (Interior), MIL-C-22750
B-6	Non-CARC Coatings (MIL-C-83286, MIL-P-14105, MIL-P-52905)
B-7	Primers (MIL-P-23377, MIL-P-53022, MIL-P-53030, MIL-P-85582)
B-8	Primer (Wash) Pretreatment (Formula No. 117 for Metals) Metric, DOD-P-15328
B-9	Remover, Paint, Epoxy, Polysulfide and Polyurethane Systems, MIL-R-81294
B-10	Thinner, MIL-T-81772
B-11	Miscellaneous

Change 3 B-1

Table B-2. Coating, Aliphatic Polyurethane, Chemical Agent Resistant (CARC)
2-Component (Topcoat) MIL-C-46168 Type II (Cont'd)

COLOR	COLOR NUMBER	NSN	SIZE
Green 383*	34094	8010-01-160-6741	1 1/4 Qt Kit
Green 383*	34094	8010-01-162-5578	1 1/4 Gal Kit
Green 383*	34094	8010-01-160-6742	5 Gal Kit
Green 383*	34094	8010-01-160-6743**	55 Gal Drum
Brown 383*	30051	8010-01-160-6744	1 1/4 Qt Kit
Brown 383*	30051	8010-01-160-6745	1 1/4 Gal Kit
Brown 383*	30051	8010-01-160-6746	5 Gal Kit
Brown 383*	30051	8010-01-160-6747**	55 Gal Drum
Dark Green	34082	8010-01-141-2412	1 1/4 Qt Kit
Dark Green	34082	8010-01-130-3343	1 1/4 Gal Kit
Dark Green	34082	8010-01-131-0611	5 Gal Kit
Dark Green	34082	8010-01-132-2977**	55 Gal Drum
Field Drab	33105	8010-01-141-2414	1 1/4 Qt Kit
Field Drab	33105	8010-01-130-3345	2 Gal Kit
Field Drab	33105	8010-01-148-3662	5 Gal Kit
Field Drab	33105	8010-01-127-8911**	55 Gal Drum
Earth Yellow	33245	8010-01-141-2415	1 1/4 Qt Kit
Earth Yellow	33245	8010-01-130-3346	1 1/4 Gal Kit
Earth Yellow	33245	8010-01-131-0612	5 Gal Kit
Earth Yellow	33245	8010-01-133-1986**	55 Gal Drum
Sand	33303	8010-01-141-2416	1 1/4 Qt Kit
Sand	33303	8010-01-130-3347	1 1/4 Gal Kit
Sand	33303	8010-01-131-6259	5 Gal Kit
Sand	33303		55 Gal Drum
Black*	37030	8010-01-141-2419	1 1/4 Qt Kit
Black*	37030	8010-01-131-6254	1 1/4 Gal Kit
Black*	37030	8010-01-131-6261	5 Gal Kit
Black*	37030		55 Gal Drum
Aircraft Green	34031	8010-01-141-2420	1 1/4 Qt Kit
Aircraft Green	34031	8010-01-131-6255	1 1/4 Gal Kit
Aircraft Green	34031	8010-01-131-6262	5 Gal Kit
Aircraft Green	34031		55 Gal Drum
Olive Drab	34088	8010-01-146-2650	1 1/4 Qt Kit
Olive Drab	34088	8010-01-055-2319	1 1/4 Gal Kit
Olive Drab	34088	8010-01-144-9875	5 Gal Kit
Aircraft Gray	36300	8010-01-144-9882	1 1/4 Qt Kit
Aircraft Gray	36300	8010-01-127-8908	1 1/4 Gal Kit
Aircraft Gray	36300	8010-01-144-9876	5 Gal Kit
Aircraft White	37875	8010-01-144-9883	1 1/4 Qt Kit

Change 3 B-2

Table B-2. Coating, Aliphatic Polyurethane, Chemical Agent Resistant (CARC)
2-Component (Topcoat) MIL-C-46168 Type II (Cont'd)

COLOR	COLOR NUMBER	NSN	SIZE
Aircraft White	37875	8010-01-144-9872	1 1/4 Gal Kit
Aircraft White	37875	8010-01-144-9877	5 Gal Kit
Aircraft Red	31136	8010-01-144-9884	1 1/4 Qt Kit
Aircraft Red	31136	8010-01-144-9873	1 1/4 Gal Kit
Aircraft Red	31136	8010-01-144-9878	5 Gal Kit
Aircraft Black	37038	8010-01-144-9885	1 1/4 Qt Kit
Aircraft Black	37038	8010-01-146-2646	1 1/4 Gal Kit
Aircraft Black	37038	8010-01-144-9879	5 Gal Kit
Interior Aircraft Black (w/Glass Beads)	37031	8010-01-144-9886	1 1/4 Qt Kit
Interior Aircraft Black (w/Glass Beads)	37031	8010-01-146-2647	1 1/4 Gal Kit
Interior Aircraft Black (w/Glass Beads)	37031	8010-01-146-4376	5 Gal Kit
Insignia Blue	35044	8010-01-144-9887	1 1/4 Qt Kit
Insignia Blue	35044	8010-01-146-2648	1 1/4 Gal Kit
Insignia Blue	35044	8010-01-144-9880	5 Gal Kit
Interior Aircraft Gray	36231	8010-01-170-7583	1 1/4 Qt Kt
Interior Aircraft Gray	36321	8010-01-146-2649	1 1/4 Gal Kit
Interior Aircraft Gray	36231	8010-01-170-0132	5 Gal Kit
Aircraft Yellow	33538	8010-01-247-8885	1 1/4 Qt Kit
Aircraft Yellow	33538	8010-01-235-8059	1 1/4 Gal Kit
Aircraft Yellow	33538	8010-01-235-5079	5 Gal Kit
Dark Sandstone	33510	8010-01-260-7480	1 1/4 Qt Kit
Dark Sandstone	33510	8010-01-160-7479	1 1/4 Gal Kit
Dark Sandstone	33510	8010-01-260-7478	5 Gal Kit
Tan	33446	8010-01-260-0910	1 1/4 Qt Kit
Tan	33446	8010-01-260-0909	1 1/4 Gal Kit
Tan	33446	8010-01-260-0908	5 Gal Kit

*Color for 3-color camouflage system

**This NSN for Component A only; Component B for all 55 Gal Drum Sizes has NSN 8010-01-132-0205. If four 55 Gal Drums of Component A are ordered, one 55 Gal Drum of Component B will also be shipped. (MIL-C-46168 is mixed 4 parts Component A with 1 part Component B.)

Table B-3. Coating, Aliphatic Polyurethane, Chemical Agent Resistant (CARC)
2-Component (Topcoat) MIL-C-46168 Type IV

COLOR	COLOR NUMBER	NSN	SIZE
Green 383*	34094	8010-01-260-7481	1 1/4 Qt Kit
Green 383*	34094	8010-01-260-0911	1 1/4 Gal Kit
Green 383*	34094	8010-01-260-0912	5 Gal Kit
Brown 383*	30051	8010-01-260-7482	1 1/4 Qt Kit
Brown 383*	30051	8010-01-260-0916	1 1/4 Gal Kit
Brown 383*	33105	8010-01-260-0917	5 Gal Kit
Field Drab	33105	8010-01-260-0918	1 1/4 Qt Kit
Field Drab	33105	8010-01-260-0919	1 1/4 Gal Kit
Field Drab	33105	8010-01-260-0920	5 Gal Kit
Sand	33303	8010-01-260-0921	1 1/4 Qt Kit
Sand	33303	8010-01-260-0922	1 1/4 Gal Kit
Sand	33303	8010-01-260-7483	5 Gal Kit
Black*	37030	8010-01-260-0913	1 1/4 Qt Kit
Black*	37030	8010-01-260-0914	1 1/4 Gal Kit
Black*	37030	8010-01-260-0915	5 Gal Kit
Tan 686	33446	8010-01-306-9680	5 Gal Kit
Tan 686	33446	8010-01-306-9681	Qt Kit
Tan 686	33446	8010-01-306-9682	5 Gal Kit

*Color for 3-color camouflage system

Change 3 B-4

Table B-4. Coating, Aliphatic Polyurethane, Chemical Agent Resistant (CARC) Single Component (Topcoat) MIL-C-53039

COLOR	COLOR NUMBER	NSN	SIZE
Green 383*	34094	8010-01-229-7546	1 Qt Can
Green 383*	34094	8010-01-229-9561	1 Gal Can
Green 383*	34094	8010-01-229-7547	5 Gal Can
Green 383*	34094	8010-01-232-8514	55 Gal Drum
Brown 383*	30051	8010-01-229-7543	1 Qt Can
Brown 383*	30051	8010-01-229-7544	1 Gal Can
Brown 383*	30051	8010-01-229-7545	5 Gal Can
Brown 383*	33051	8010-01-233-0600	55 Gal Drum
Black*	37030	8010-01-229-7540	1 Qt Can
Black*	37030	8010-01-229-7541	1 Gal Can
Black*	37030	8010-01-229-7542	5 Gal Can
Black*	37030	8010-01-233-1568	55 Gal Drum
Sand	33303	8010-01-234-2934	1 Qt Can
Sand	33303	8010-01-234-2935	1 Gal Can
Sand	33303	8010-01-234-2936	5 Gal Can
Sand	33303		55 Gal Drum
Aircraft Green	34031	8010-01-246-0717	1 Qt Can
Aircraft Green	34031	8010-01-246-0718	1 Gal Can
Aircraft Green	34031	8010-01-246-0719	5 Gal Can
Aircraft Green	34031	8010-01-246-0255	55 Gal Drum
Tan 686	33446	8010-01-276-3638	Qt
Tan 686	33446	8010-01-276-3639	Gal
Tan 686	33446	8010-01-276-3640	5 Gal
Tan 686	33446	8010-01-276-3641	55 Gal
Aircraft Red	31136	8010-01-254-5850	Gal
Aircraft Black	37038	8010-01-254-8444	Gal

*Color for 3-color camouflage system

Change 3 B-5

Table B-5. Epoxy Coating (Interior) MIL-C-22750

COLOR	COLOR NUMBER	NSN	SIZE
Gray	16473	8010-01-350-2072	2 Qt Kit
Gray	36099	8010-01-350-6253	2 Qt Kit
Gray, Lt	36495	8010-01-314-6066	2 Qt Kit
Gray, Lt.	36495	8010-01-314-6067	2 Gal Kit
Gray	36320	8010-01-316-3034	2 Qt Kit
Gray	36320	8010-01-316-3035	2 Gal Kit
Gray	36375	8010-01-117-7689	2 Qt Kit
Gray	36375	8010-01-316-3039	2 Gal Kit
Gull Gray	16440	8010-01-313-8119	2 Qt Kit
Gray	36440	8010-01-316-3043	2 Gal Kit
Gray	36231	8010-01-316-2551	2 Gal Kit
Gray	26492	8010-01-350-2679	2 Gal Kit
Gray	26492	8010-01-350-2679	2 Gal Kit
Gray	26622		1 Gal Kit
Gray	26081	8010-01-350-2074	2 Gal Kit
Gray	16081	8010-01-350-2071	2 Gal Kit
Gray	36231	8010-01-316-2550	2 Qt Kit
Gray	36231	8010-01-316-2550	2 Qt Kit
Orange-Yellow	13538	8010-01-313-7292	2 Qt Kit
Yellow	13538	8010-00-148-3166*	2 Gal Kit
Yellow	23538	8010-01-350-4735	2 Gal Kit
Yellow	13538	8010-01-313-8110	2 Gal Kit
White	17925	8010-01-313-8700	2 Qt Kit
White	17925	8010-01-313-8700	2 Qt Kit
White	17925	8010-00-082-2439	2 Gal Kit
White	27875	8010-01-350-4733	2 Gal Kit
White	17925	8010-01-314-4497	10 Gal Kit
Insignia Blue	15044	8010-01-314-2524	2 Qt Kit
Insignia Blue	35044	8010-01-350-4732	2 Qt Kit
Blue	35237	8010-01-118-9981	2 Qt Kit
Blue	35237	8010-01-314-4704	2 Gal Kit
Blue	35299		2 Gal Kit
Maroon	10049	8010-01-350-4729	2 Qt Kit
Insignia Red	11136	8010-01-053-2649	2 Qt Kit
Red	31302	8010-01-350-6255	2 Gal Kit
Orange-Red	12199	8010-01-350-5241	2 Qt Kit
International Orange	12197	8010-01-313-7288	2 Qt Kit
International Orange	12197	8010-00-948-6733	2 Gal Kit
Olive Drab	34088	8010-01-350-2070	2 Gal Kit
Olive Drab	24084	8010-01-350-5240	2 Gal Kit
Olive Drab	24084	8010-01-350-5240	2 Gal Kit
Olive Drab	34088	8010-01-350-2070	2 Gal Kit
Clear	Full Gloss	8010-01-313-8702	1 Qt Kit
Aluminum	Full Gloss		1 Pt Kit

*Not Listed on AMDF; must be specially ordered.

Table B-5. Epoxy Coating (Interior) MIL-C-22750 (Cont'd)

COLOR	COLOR NUMBER	NSN	SIZE
Black	37038	8010-01-314-6071	2 Qt Kit
Black	37038	8010-01-314-6072	2 Gal Kit
Dark Green	14062	8010-01-350-4730	2 Qt Kit
Green	24052	8010-01-350-2678	2 Gal Kit
U Green	14187	8010-01-313-7293	2 Qt Kit
Seafoam Green	24533	8010-01-211-9645	2 Qt Kit
Seafoam Green	24533	8010-01-212-1710	2 Gal Kit
Seafoam Green	24533	8010-01-314-2528	10 Gal Kit

*Not listed on AMDF; must be specially ordered.

Table B-6. Non-CARC Coatings

COATING	MILSPEC	COLOR	COLOR NUMBER	NSN	SIZE
Coating, Urethane,	MIL-C-83286	Yellow	34079	8010-00-181-8297	2 Qt Kit
Aliphatic Isocynate	MIL-C-83286	Yellow	13538	8010-00-181-8292	2 Gal Kit
For					
Aerospace	MIL-C-83286	Yellow	33538	8010-00-181-8302	2 Qt Kit
Applications					
(Contains Lead)	MIL-C-83286	Yellow	33538	8010-00-181-8300	2 Gal Kit
Paint, Heat	MIL-P-14105	Green	34094	8010-01-235-2693	1 Qt Can
Resistant (For					
Use on Surfaces	MIL-P-14105	Green	34094	8010-01-235-4164	1 Gal Can
Exceeding					
400°F (204°C)	MIL-P-14105	Brown	30051	8010-01-235-2694	1 Qt Can
	MIL-P-14105	Brown	30051	8010-01-235-2695	1 Gal Can
	MIL-P-14105	Black	37030	8010-01-235-4165	1 Qt Can
	MIL-P-14105	Black	37030	8010-01-235-4166	1 Gal Can
Paint, Arctic	MIL-P-52905	White			1.Gal Can
Camouglage,					
Removable (White)					

Change 3 B-7

Table B-7. Primers

PRIMER	MILSPEC	TYPE	COLOR	COLOR NUMBER	NSN	SIZE
Primer Coating, Epoxy-Polyamine	MIL-P-23377	I	Deep Yellow	-	8010-00-229-4813	1 Pt Kit
Chemical and Solvent Resistant (Contains	MIL-P-23377	I	Deep Yellow	-	8010-00-142-9279	1 Qt Kit
Chromate for use on Non-ferrous	MIL-P-23377	I	Deep Yellow	-	8010-00-935-7080	2 Qt Kit
Surfaces)	MIL-P-23377	I	Yellow		8010-00-082-2450	2 Gal Kit
	MIL-P-23377	II	Dark Green	34052	8010-01-048-6539 8010-00-082-2477	2 Gal Kit
	MIL-P-23377	I	Deep Yellow			10 Gal Kit
Primer, Epoxy Coating (Corrosion	MIL-P-53022	-	White	-	8010-01-193-0516	1 1/4 Qt Kit
Inhibiting - For Use	MIL-P-53022	-	White	-	8010-01-193-0517	1 1/4 Gal Kit
Ferrous and Non-ferrous Surfaces)	MIL-P-53022	-	White	-	8010-01-187-9820	5 Gal Kit
Primer Coating, Epoxy, Water	MIL-P-53030	-	Reddish Brown	-	8010-01-193-0519	1 1/4 Qt Kit
Reducible (For Use on Ferrous and	MIL-P-53030	-	Reddish Brown	-	8010-01-193-0520	1 1/4 Gal Kit
Non-ferrous Surfaces) Brown	MIL-P-53030	-	Reddish	-	8010-01-193-0521	5 Gal Kit
Primer Coating: Epoxy, VOC Compliant	MIL-P-85582	-	Light Green	-	8010-01-218-0856	1 Qt Kit
Chemical and Solvent Resistant (Lead-Free, Water Reducible	MIL-P-85582	-	Light Green	-	8010-01-218-7354	1 Gal Kit

Change 3 B-8

Table B-8. PRIMER (Wash) PRETREATMENT
(Formula No. 117 for Metals)
METRIC, DOD-P-15328

NSN	SIZE
8030-00-850-7076	1.25 Qt Kit
8030-00-281-2726	1gal kit
8030-00-165-8577	5 Gal Kit

Table B-9. REMOVER, PAINT, EPOXY,
POLYSULFIDE AND POLYURETHANE
SYSTEMS, MIL-R-81294

TYPE	NSN	SIZE
I	8010-00-142-9273	1 Pt
I	8010-00-181-7568	1 Gal
I	8010-00-926-1488	5 Gal
I	8010-00-926-1489	55 Gal Drum

Table B-10. THINNER, MIL-T-81772

TYPE	NSN	SIZE
I (Polyurethane)	8010-00-181-8080	1 Gal Can
I (Polyurethane)	8010-00-181-8079	5 Gal Can
I (Polyurethane)	8010-00-280-1751	55 Gal Drum
II (Epoxy)	8010-01-200-2637	1 Gal Can
II (Epoxy)	8010-01-212-1704	5 Gal Can
II (Epoxy)	8010-01-168-0684	55 Gal Drum

Table B-11. MISCELLANEOUS

ITEM	SIZE/TYPE	NSN	P/N	CAGEC
Viscosimeter, Cup	No. 2		VG-8202	96173

Change 3 B-9

Table B-12. WOOD SEALERS

TYPE	NSN	SIZE
1 -Component	8010-01-327-6479	5 Gallon
1 -Component	8010-01-327-6480	55 Gallon
2-Component	8010-01-327-5190	10 Gallon
2-Component	8010-01-327-5189	110 Gallon

APPENDIX C
COLORS FOR ARMY MATERIEL

C-1. This appendix contains guidelines for colors used in painting Army materiel. Tables C-2 through C-9 cover the various types of Army equipment. Refer to Table C-1 to find which table covers a given equipment type.

Table C-1. Color Tables.

Table Number	Equipment Type
C-2	Amphibians and Vessels
C-3	Army Aircraft and Surveillance Drones
C-4	Railroad Equipment
C-5	Vehicles, Construction Equipment, and Materials Handling Equipment
C-6	Missiles, Heavy Rockets, and Related Missile Ground Support Equipment
C-7	Communications-Electronic Equipment
C-8	Bridging Equipment
C-9	Other Equipment

Table C-2. Amphibians and Vessels

ITEM NO	EQUIPMENT	COLOR	PLACEMENT
1.	Mobile floating assault bridge (MAB) transporter and superstructure	Camouflage paint pattern using MIL-C-46168 or MIL-C-53039 Aircraft white 37875 using MIL-C-22750	Exterior Interior ferrous components
2.	Lighter, Air Cushion Vehicle (LACV-30)	Not to be painted unless deemed necessary by operating command.	
3.	Lighter, Amphibious Resupply Cargo (LARC) LX	Camouflage paint pattern using MIL-C-46168 or MIL-C-53039. **NOTE** **Markings may be adhesive-backed markers or paint conforming to color requirements.**	Exterior
4.	Unit Identification	Lusterless Black 37030	
5.	Other	See TB 43-0144	See TB 43-0144

Table C-3. Army Aircraft and Surveillance Drones

ITEM NO	EQUIPMENT	COLOR	PLACEMENT
1.	Army aircraft andSee surveillance drones	TM 55-1500-345-23 andSee TM for specific aircraft	TM 55-1500-345-23 and TM for specific aircraft

Table C-4. Railroad Equipment

ITEM NO	EQUIPMENT	COLOR	PLACEMENT
1.	Locomotive, tenders, and work equipment	Semigloss Black 27038	Exposed exterior surfaces
2.	Rolling stock		
	a. Freight passenger, hospital, kitchen and caboose cars	Gloss Olive Green	Exposed exterior surfaces
	b. Gondola, hopper, and tank cars	Gloss Black 17038	Exposed exterior surfaces
		NOTE **Markings may be adhesive backed markers or paint conforming to color requirements**	
3.	Agency identification, "US Army"	Gloss White 17875	On both sides of all railroad equipment; exception in theaters of operation, do not apply to freight cars and cabooses
4.	Identification Numbers	Gloss White 17875	a. On both sides and each end of all locomotives, locomotive tender units, work equipment, and roling stock b. On rolling stock, car numbers will be prefixed with letters "USA" immediately to the left of or above car number

Table C-4. Railroad Equipment-Continued

ITEM NO	EQUIPMENT	COLOR	PLACEMENT
4.	Identification Numbers (cont'd)	Gloss White 17875	c. In CONUS, freight and caboose cars acceptable for interchange on commercial railroads will use the prefix "USAX" in lieu of the above where specifically authorized by the Director of Supply, USAMC. In such cases the identification number and "USAX" prefix also will be marked on the ends of the car.
5.	Army Medical Department markings:		
	a. Red Cross insignia	Gloss Red 11136 and Gloss White 17875	On both sides and on the roof of all kitchen, baggage, ambulance ward, and personnel cars assigned Army Medical Department. The insignia applied to the side of the cars will be located near the end of the car.
	b. "ARMY MEDICAL DEPARTMENT"	Gloss White 17875	These words and type of car will be marked on both sides of Army Medical Department car near the center of the car. The type of car also will be marked in conjunction with identification number of the car.
	c. Army Medical Department insignia (red caduceus on a circular white field)	Gloss Red 11136 and Gloss White 17875	On both sides of Army Medical Department cars near the center of the car. Place on each side of or or adjacent to the words, ArMY MEDICAL DEPARTMENT", and type of car marked as in b. above.

Table C4.   Railroad Equipment-Continued

ITEM NO	EQUIPMENT	COLOR	PLACEMENT
6.	Safety markings (CONUS only)	Gloss Yellow 13538 or reflective yellow adhesive backed marker conforming to Federal Specification L-S-300, color j. **NOTE** **All colors and markings for railroad equipment to conform to MIL-P-53044.**	A stripe will be marked across each side and across each and of all locomotive tender units. Access steps to locomotive and locomotive tenders also will be painted.

Table C-5.   Vehicles, Construction Equipment, and Materials Handling Equipment

ITEM NO	EQUIPMENT	COLOR	PLACEMENT
1.	Vehicles, construction equipment, and materials handling equipment from new production, depot reconditioning, or depot stocks; exceptions listed below.	Three-color camouflage pattern using MIL-C-46168 or MIL-C-53039	Exterior surface unless otherwise directed
		Lusterless Green 383	Interior surfaces unless otherwise directed
		Walkway Compound (MIL-W-5044)	Under topcoat over walking areas (catwalks, walkways, platforms, cabs, fenders, frames, guards, foot-operated controls, mounting steps, ladders, top bumpers, slope and deck plates, turret floors, crew compartments, ramps, van floors, etc.), and interior surfaces
2.	Firefighting vehicles auxiliary vehicles assigned exclusively for firefighting or protection against fires, and air field crash rescue vehicles	Gloss Red 11136 or Gloss Yellow 13670 when in non-tactical use; in hot climates, Gloss White 17875 may be used.	Exterior and interior surfaces

Table C-5. Vehicles, Construction Equipment, and Materials Handling Equipment-Continued

ITEM NO	EQUIPMENT	COLOR	PLACEMENT
3.	Repair and utility vehicles including attachments used for highway construction or maintenance (excluding construction equipment assigned to tactical units)	Gloss Yellow 13538	Exterior and interior surfaces
4.	a. Materials handling equipment, including air-craft towing, and fuel and oil dispensing vehicles used in nontactical areas	Gloss Yellow 13538 surfaces	Exterior and interior
	b. Materials handling equipment, including fuel and oil dispensing equip-ment used in tactical areas	Three-color camouflage pat-tern using MIL-C-46168 or MIL-C-53039	Exterior surfaces
		Lusterless Green 383	Interior surfaces
5.	Garbage and refuse collec-tion trucks in nontactical areas	Gloss White 17875 and Gloss Black 17038	Exterior and interior sur-faces; top of hood handles, and control levers, and stepping areas of running board. Stepping area may be covered with black non-slip walkway compound MIL-W-5044, type IV.
6.	Commercial design vehicles in administrative use (Bodies on vehicles need not be painted if con-structed of aluminum, stainless steel or other cor-rosive resistant material, and when ordinarily not painted in common com-mercial practice)	Gloss Green 14050 except where otherwise provided for special type vehicles assigned to special use. In tactical use, colors will con-form to 3-color camouflage pattern using MIL-C-46168 or MIL-C-53039.	Exterior and interior surfaces

Table C-5. Vehicles, Construction Equipment, and Materials Handling Equipment-Continued

ITEM NO	EQUIPMENT	COLOR	PLACEMENT
7.	Vehicles and equipment used in connection with approved research and development test projects	Any color deemed appropriate by the head of the responsible activity	Exterior and interior surfaces
8.	Calibration vehicles in CONUS tactical environment, 3-color camouflage pattern using MIL-C-46168 or MIL-C-53039. Otherwise, solar heat-reflecting Semigloss Olive Drab Enamel 24087 or Gloss White 17875 conforming to MIL-E-46136.	In overseas commands where vehicles may be in	Exterior surfaces
9.	Refrigerator vans in non-tactical use in hot climates; water tank vehicles in non-tactical use in hot climates; van type vehicles in non-tactical use in hot climates assigned to centers or schools for the purpose of training personnel in the operation of equipment installed therein; van type vehicles and shelters in non-tactical use in hot climates in which installed equipment would be adversely affected, or personnel would not be able to accomplish assigned technical functions due to excessive heat from solar radiation.	Semigloss Green 24533 Lusterless Gray 36118 Gloss White 17875 or solar heat-reflecting Semigloss Olive Drab Enamel 24087 conforming to MIL-E-46136 or solar heat-reflecting Lusterless Olive Drab Enamel conforming to MIL-E-46117	Walls, ceiling, doors, fittings, and mounted equipment Floors Applicable exterior surfaces, upon authorization by the theater commander
10.	a. Tracked combat vehicles and special purpose armored hull-type vehicles. Exceptions: b and c below	Three-color camouflage pattern using MIL-C-46168 or MIL-C-53039.	Exterior and surfaces which become exterior surfaces during use(i.e. doors, hatches)

Table C-5. Vehicles, Construction Equipment, and Materials Handling Equipment-Continued

ITEM NO	EQUIPMENT	COLOR	PLACEMENT
		White conforming to MIL-C-22750	Interior
		Walkway Compound MIL-W-5044, Type IV	Walkways, floors, ramps
	b. MI113 personnel carrier family of vehicles, and the M2 and M3 Bradley Fighting vehicles	Three-color camouflage pattern using MIL-C-46168 or MIL-C-53039	Exterior surfaces and surfaces (hatches, ramps,etc.) which become exterior surfaces during operational use
		Seafoam Green (Color 24533) Epoxy MIL-C-22750	Interior surfaces
		Walkway compound MIL-W-5044	Floors (under MIL-C-22750 topcoat)
	c. Vehicles and construction equipment in nontactical use, in Army units not subject to the Army camouflage policy, not already camouflaged with three color MIL-C-46168 or MIL-C-53039 and not identified in items 2, 3, 4a, 4b, 5, 6, 7, 8, 9, 12, 13, 14, and 16	Lusterless Green 383, 34094, MIL-C-46168 or MIL-C-53039	Exterior and interior surfaces unless otherwise directed
11.	Interior surfaces of van type bodies equipped with interior lighting in which personnel are required to perform certain detail operations	Seafoam Green Epoxy MIL-C-22750, color 24533	Interior surfaces
		Walkway Compound MIL-W-5044	Floors, under MIL-C-22750 topcoat
12.	Commercial design vehicles assigned to Armed Forces Policy Units	Gloss Black 17038	Exterior and interior surfaces
13.	Commercial design vehicles used for military police traffic accident prevention	Gloss White 17875 when deemed necessary by the responsible commander	Exterior and interior surfaces

Table C-5. Vehicles, Construction Equipment, and Materials Handling Equipment-Continued

ITEM NO	EQUIPMENT	COLOR	PLACEMENT
14.	Vehicles used for intelligence, criminal investigation, and similar purposes, requiring concealment of the true identity of the activity involved	Painted and marked as deemed appropriate by the responsible commander	Exterior and interior surfaces **NOTE** **Proper identification of such vehicles will be marked in an inconspicuous location as prescribed in applicable technical publications and AR 381-143.**
15.	Ground support equipment in Army units subject to the Army camouflage policy	3-color camouflage pattern using MIL-C-46168 or MIL-C-53039	Exterior surfaces
16.	Ambulances in tactical use	3-color camouflage pattern using MIL-C-46168 or MIL-C-53039 Seafoam Green, MIL-C-22750, color 24533 Lusterless Black 37030, MIL-C-46168 or MIL-C-53039	Exterior surfaces, and surfaces (doors, etc.) which become exterior surfaces during operational use Interior surfaces Interior of cab when vehicle is commercial design procured with standard color options
17.	National symbol (5-pointed star)	Star will be a size to fit into a 3" diameter circle in Lusterless Black 37030, MIL-C-46168 or MIL-C-53039	Star will be applied at time of manufacture or by depots only when material is specifically designated for issue to US Army units.

Table C-5. Vehicles, Construction Equipment, and Materials Handling Equipment-Continued

ITEM NO	EQUIPMENT	COLOR	PLACEMENT
	a. Tactical and combat front vehicles and related equipment, including support and special purpose vehicles in CONUS and overseas	Lusterless Black 37030, MIL-C-46168 or MIL-C-53039	One each, centered on and rear, on any suitable vertical or near vertical surface such as bumper or tailgate, where it will not be obscured by canvas items, gasoline cans, pioneer tools or other objects. Normally, stars will not be applied to ambulances or other medical dedicated vehicles, or to ground support equipment at Army airfields in CONUS.
	b. Mounted equipment	Lusterless Black 37030, MIL-C-46168 or MIL-C-53039	Wherever there is suitable surface on the vehicle
18.	Unit identification (abbreviations representing the complete identification of the unit to which the vehicle or equipment is assigned). Specific abbreviations to be used, and methods of application are prescribed in TB 43-0209. (See c and d of note at end of this item.)	Lusterless Black 37030, MIL-C-46168 or MIL-C-53039	a. Markings will be in uniform Gothic style letters, the numerals to be thelargest size practical for use in available space, but not to exceed 4 inches in height. b. Normally markings will appear on both front and rear of each item of equipment; usually on the bumpers when so equipped. Where a more suitable surface is available, such surface may be used provided the location is not in conflict with the location of other prescribed markings. Where a suitable surface is not available on the front or rear of equipment, use an appropriate location on the sides of the item.

Table C-5. Vehicles, Construction Equipment, and Materials Handling Equipment-Continued

ITEM NO	EQUIPMENT	COLOR	PLACEMENT
			c. Markings will be applied to vehicles and equipment by the organization to which material is assigned. Such markings will be maintained on vehicles and equipment at all times, but will be removed when the item is permanently transferred from the operating unit. d. When directed by the responsible Commander for security purposes, the first two categories of markings (see a and b of footnote below) will be thoroughly removed from all vehicles and equipment being shipped with units from CONUS to overseas commands. To insure effective removal, the original figures must be physically erased rather than painted over.

NOTE

Unit Identification markings are divided into four elements arranged from left to right, as follows:

a. Major command, organization, or activity. The major headquarters having jurisdiction over the vehicle or equipment, normally not lower than a division, brigade, group, or major subordinate command.

b. Intermediate organization or activity. The next lower headquarters having jurisdiction over the vehicle or equipment, normally the headquarters above the unit to which the vehicle or equipment is assigned. This includes regiments, separate battalions, installations, and separate companies or detachments not assigned to an intermediate headquarters.

c. Unit or activity. The lowest unit or activity to which the vehicle is assigned, normally a company type unit. This space may be used to identify the type of separate company or detachment already identified above.

d. Vehicle or equipment number. The sequence number of the vehicle or equipment in the normal order of march within the unit to which it is assigned.

Table C-5. Vehicles, Construction Equipment, and Materials Handling Equipment-Continued

ITEM NO	EQUIPMENT	COLOR	PLACEMENT
19.	* Agency identification and registration number:		
	a. Vehicles and construction equipment	Colors specified on 3-color camouflage drawings, but constrasting with color patch painted over.	Exterior surfaces
	b. Agency identification; e.g., "US ARMY", will be painted on those vehicles, construction equipment and materials handling equipment in nontactical use, in Army units not subject to the Army camouflage policy, and vehicles and equipment as identified in items 2, 3, 4a, 5, 6, 7, 8, 9, 10c, 12, and 13	Lusterless Black 37030, MIL-C-46168 or MIL-C-53039 practical for use in available space, but not to exceed 4 inches in height. Normally agency identification will appear on both sides of each item of equipment.	Marking will be in uniform Gothic style letters, the letters to be the largest size
	c. Registration number assigned in accordance with AR 708-1 will be carried by those vehicles construction equipment, and materials handling equipment which they carry agency identification marking (e.g., US ARMY), in accordance with criteria set forth in b above.	Lusterless Black 37030, MIL-C-46168 or MIL-C-53039	Exterior: Both sides an rear in uniform Gothic letters, no larger than 4 inches high.
	d. Mounted equipment MIL-C-46168 or MIL-C-53039	Lusterless Black 37030, necessary.	May be applied when

* The identification and registration number shall be placed on any appropriate interior area which is visible from outside a locked or secured item.

Table C-5. Vehicles, Construction Equipment, and Materials Handling Equipment-Continued

ITEM NO	EQUIPMENT	COLOR	PLACEMENT
20.	General officer vehicle identification		
	a. Plate	Gloss enamel, color in accordance with AR 840-10 and MIL-P-14631	Right end of the front bumper. If plate interferes with lights or functional components of the vehicle, mount in front center. Remove or cover when general officer is not riding in vehicle.
	b. Flag	See AR 840-10	
21.	Weight classification for bridge crossing	Black numerals on a yellow circular background 9 inches in diameter. When towing another vehicle, the weight classification number of the combination will be shown with the letter "C" in red above the number.	On the front of applicable self-propelled vehicles For design of plate and mounting, see TB 43-0209.
	a. All vehicles with a gross weight of over 3 tons; all trailers with a rated payload of 1-1/2 tons and over	On vehicles painted MIL-C-46168 or MIL-C-53039 lusterless colors in accordance with camouflage policy, numbers 3 inches high will be painted in contrasting colors directly over camouflage pattern. On vehicles painted in gloss colors, use Gloss Yellow 13538 or Gloss Red 11136. On vehicles painted in semigloss or lusterless colors, use Lusterless Yellow 33538, Gloss Black 17038, and Gloss Red 11136.	

Table C-5. Vehicles, Construction Equipment, and Materials Handling Equipment-Continued

ITEM NO	EQUIPMENT	COLOR	PLACEMENT
	b. Combination vehicles	Black (17038) numerals on a yellow (33538) circular background 6 inches in diameter	The gross weight classification of the prime mover alone and of the towed vehicle alone, will be marked on the right side of the vehicles.
22.	Safety.		
	a. Stripes (vehicles and equipment in nontactical use, which because of size, construction, or function present a possible hazard)	Alternate Gloss Yellow 13655 and Gloss Black 17038 stripes in accordance with AR 385-30	Normally, stripes are applied only to the rear, but also may be applied to the front and certain side surfaces when the conditions warrant.
	b. "FLAMMABLE" and "NO SMOKING WITHIN 50 FEET"- vehicles used for bulk transportation of gasoline, fuel oil, or other flammable liquids by Army units subject to the Army camouflage policy	Brown 383 over green and black. Lusterless Black 37030 or 37038 over brown. The word "FLAMMABLE" will be in 6 inch letters; the words "NO SMOKING WITHIN 50 FEET" will be in 3 inch letters.	On both sides and rear of body, "NO SMOKING WITHIN 50 FEET" should be on a line below "FLAMMABLE".
	"FLAMMABLE" and "NO SMOKING WITHIN 50 FEET"- vehicles used for bulk transportation of gasoline, fuel oil, or other flammable liquids by Army units not subject to the Army camouflage policy, or in nontactical use	When vehicle is commonly used on public highways in CONUS, Title 49, Sec. 177.823, Code of Federal Regulations, requires 6 inch block letters in Gloss Red 11105 or 11136 for word "FLAMMABLE" and 3 inch block letters "NO SMOKING WITHIN 50 FEET" also in Gloss Red 11105 or 11136. Gloss White 17875 will be used as background extending one inch all around the lettering. If available, reflective markings should be used.	On both sides and rear of body. "NO SMOKING WITHIN 50 FEET" should be on a line below "FLAMMABLE". Vehicles used infrequently on public highways must employ removable or reversible signs which are securely fastened while in use. These safety markings will be applied upon receipt by the using service.

Table C-5. Vehicles, Construction Equipment, and Materials Handling Equipment-Continued

ITEM NO	EQUIPMENT	COLOR	PLACEMENT
	c. "GALS CAP" - Fuel and oil dispensing vehicles and equipment used for servicing aircraft		On side of tank near each manhole cover. In addition, type of fuel or oil by military designation will be clearly marked.
	d. Placards for vehicles carrying explosive or other dangerous articles		
	(1) Over public highways		Placards will be used in accordance with applicable DOT regulations and AR 55-355.
	(2) In training areas		Placards will be used in accordance with applicable DOT regulations and AR 385-63.
	e. Reflective markings (vehicles and ground support equipment used on airfield flight lines)		As Specified in TM 55-1500-204-25/1
	f. Flags (all vehicles not painted in accordance with items 2 and 9 of this table and e above).		Vehicles will bear a distinctive flag whenever operating on landing areas, runways, taxiways, or peripheral roads at airfields. The flag will be square, at least 3 feet on each side, and will be divided into 9 equal size squares forming a checkerboard pattern with the center and the corner squares in international orange and the remaining 4 squares in white.
			A red flag will be mounted on vehicles to indicate danger when considered necessary to caution personnel of a hazardous condition in the area.

Table C-5. Vehicles, Construction Equipment, and Materials Handling Equipment-Continued

ITEM NO	EQUIPMENT	COLOR	PLACEMENT
	g. DA Label 76 (A Good Driver), to promote good driving		Display in all motor vehicles to be readily visible to the driver. Requisition through normal publication channels.
23.	Red Cross insignia	Aircraft Red 31136 on Aircraft White 37875, MIL-C-46168 or MIL-C-53039	See table C-4, item 5. On tactical ambulances, cross will be approximately 36" to 48" square and will be placed on roof, both sides, rear and front if practicable. On other medical dedicated TOE vehicles, a 3" square cross will be placed on front and rear in lieu of the national symbol.
24.	Military police and security guard; tactical type vehicles:		
	a. "MILITARY POLICE"	Gloss Black 17038 on Aircraft White 37875 (MIL-C-46168 or MIL-C-53039) black letters on white background	As specified in applicable DA technical publications. On 1/4 ton military trucks, a horizontal strip with lettering will be placed under the windshield and on a disc covering the spare wheel.
	b. "SECURITY POLICE" (at class II installations and activities)	Same as above, except the words "SECURITY POLICE" will be used in lieu of "MILITARY POLICE ".	Same as above
25.	Rifle and pistol team vehicles (decals of approved design)		On both sides of vehicles
26.	Calibration service vehicles		In accordance with applicable DA technical publications
27.	Convoy markings (as prescribed by FM 55-30)	Serial letter or number may be black on reflective signboard background.	

Table C-5. Vehicles, Construction Equipment, and Materials Handling Equipment-Continued

ITEM NO	EQUIPMENT	COLOR	PLACEMENT
28.	Tactical markings (may consist of stripes, geometrical figures, other simple designs, or naming of individual vehicle). The markings will be used to provide a more visible means of identifying the vehicles of tactical units by personnel of those units during tactical operations. They will be of a suitable size to facilitate identification from ground or from a reasonable distance. They will be designed to be as nearly unintelligible as possible to enemy observation. Under no circumstances will the markings represent the numerical designation or distinctive insignia of any unit. Under actual combat conditions, no written record will be made, nor published system of tactical markings used.	Any appropriate color	As prescribed by commanders of major Army commands; any changes must be approved by these commanders.
29.	Priority signs (The signs will be authorized by the area commander. They are valid only within the area under his jurisdiction. Signs must not be displayed when vehicle is not actually being used for a priority mission and must be removed when no longer required for that mission.)		On front and rear of vehicle. They will be fabricated and mounted on vehicles in accordance with TB 43-0209.

Table C-5. Vehicles, Construction Equipment, and Materials Handling Equipment-Continued

ITEM NO	EQUIPMENT	COLOR	PLACEMENT
30.	Air recognition panels	Red fluorescent on one side and yellow fluorescent on the reverse side	Marked as directed by the appropriate major commander as to the arrangements and conditions under which they will be used. They will be draped on vehicle in a standard unchanging pattern different from displays prescribed for other recognition purposes, such as front lines, targets, etc.. Size: 6' X2'3", each panel having a tie cord of adequate length in each corner.
31.	Commercial design vehicle markings (unless otherwise specified, vehicles assigned to TOE or TD units, or to other organizations and activities for nonadministrative use)		Similar to markings used on military design vehicles

NOTE

For additional information on painting and marking of military vehicles, construction equipment and materials handling equipment, refer to TB 43-0209.

Table C-6. Missiles, Heavy Rockets, and Related Ground Support Equipment for Missile Systems

ITEM NO	EQUIPMENT	COLOR	PLACEMENT
1.	Missiles, heavy rockets and related ground support equipment and containers.	Camouflage paint pattern in accordance with applicable pattern drawing.	Exterior surfaces.
		On items for which pattern drawings have not been developed, green 383, color #34094.	Exterior surfaces.
		On items requiring desert color, tan 686, color #33446.	Exterior surfaces.
2.	Surfaces thru which RF energy passes, including but not limited to: radars, radomes, antenna, and radiating elements and covers.	Not painted unless specifically authorized by pattern drawing or applicable TM.	
3.	Vehicles (wheeled & tracked), including materials handling, personnel carrying, special purpose armored hull-type, and fighting.	As set forth in Table C-5 unless specified in applicable TM.	Exterior and interior surfaces.
4.	Electronic-Communication Shelters	As set forth in Table C-7, unless specified in applicable TM.	Exterior and interior surfaces.
5.	Markings		
	a. Agency identification (US ARMY)	For camouflage patterns, lusterless black, color #37030, on green or brown background; and, green 383, color #34094, on black background. On all other backgrounds, lusterless black, color #37030.	As specified in the applicable TM, or marked on at least two of the missile or rocket components. Placement will be along the longitudinal axis so that when the missile is in the horizontal position, the letters will be vertical.
	b. Other. (See notes.)	Same as 5.a.	As specified in applicable TM.

NOTES

(1) Paints on exterior surfaces shall conform to MIL-C-46168 or MIL-C-53039.
(2) Paints on interior surfaces shall conform to MIL-C-22750.
(3) Marking colors shall conform to the color requirements of 5.a., except for special requirements, e.g., ammunition, transportation, safety, etc.
(4) Markings shall be made using the same type paint used for painting.

Table C-7. Communications-Electronic Equipment[1]

ITEM NO	EQUIPMENT	COLOR	PLACEMENT
1.	Photographic and audio-visual	Lusterless Green 383 (34094) or Black (37030) MIL-C-46168 or MIL-C-53039 as appropriate	Exterior surfaces
2.	Tactical communications (see TB 750-10)	Lusterless Green 383, MIL-C-46168 or MIL-C-53039	
3.	Avionics	Lusterless Green 383 (34094) or Black (37030) MIL-C-46168 or MIL-C-53039 as appropriate	
4.	Meteorological	Lusterless Green 383, MIL-C-46168 or MIL-C-53039	
5.	Radiac	Same as above	
6.	Marine communication and electronics	Same as above, except as noted in 7 and 16 below	
7.	Antenna towers for use under nontactical conditions	Alternating bands of Gloss Orange 12197 and Gloss White 17875 with the band at each end colored orange	As specified in DOT's AC 70/7460-1, Obstruction, Marking & Lighting, the bands should be of equal width with each band approximately one-seventh the height of the tower. Each band will have a width of 1-1/2 to 40 feet.
8.	Armed Forces Radio and Television Network	Lusterless Green 383 (34094) or Aircraft Gray (36300), MIL-C-46168 or MIL-C-53039	
9.	Antenna Radomes	Lusterless Green 383 (34094) or Aircraft White (37875), MIL-C-46168 or MIL-C-53039	**CAUTION** **See NOTE 2 before applying any paint.**
10.	Automatic data processing	Lusterless Green 383 (34094) or Aircraft Gray (36300), MIL-C-46168 or MIL-C-53039	

Table C-7. Communications-Electronic Equipment-Continued

ITEM NO	EQUIPMENT	COLOR	PLACEMENT
11.	Ground controlled approach (GCA) radar equipment used at Army airfields:		
	a. Tactical equipment authorized by TA/TOE	Lusterless Green 383 (34094), MIL-C-46168 or MIL-C-53039, including wave guides & antenna reflectors	**CAUTION** **See NOTE 2 before applying any paint.**
	b. Equipment authorized for CONUS Army airfields by TA	In accordance with Federal Aviation Administration Technical Standards Order, TSON 26	
12.	Communications equipment installed at fixed facilities such as radio station equipment, telephone control office sets	Lusterless Green 383 (34094) or Aircraft Gray (36300), MIL-C-46168 or MIL-C-53039	
13.	Electronics test equipment	a. For bench use, semigloss Gray 26307 or Aircraft Yellow (33538) b. For field use, Lusterless Green 383 (34094) or Aircraft Gray (36300), MIL-C-46168 or MIL-C-53039	
14.	COMSEC (FSN 5810)	See TB 750-10	Communication Security (COMSEC) equipment used by the Department of the Army but design controlled by the National Security Agency (NSA), will be painted or marked or both in accordance with NSA's requirements.
15.	Special intelligence	See AR 381-143 equipment	
16.	Antenna or radiating elements, antenna base insulators or fiberglass casing	Will not be painted unless special approval is obtained from the design activity	**CAUTION** **See NOTE 2 before applying any paint.**

Table C-7. Communications-Electronic Equipment-Continued

ITEM NO	EQUIPMENT	COLOR	PLACEMENT
17.	Electronic-communication shelters (see TB 43-0118):		
	a. Configured (Shelter includes internally mounted equipment)	Three-color MIL-C-46168 or MIL-C-53039 camouflage pattern	Exterior surfaces and doors which may become exterior surfaces during operational use
		Semigloss Green 24533 conforming to MIL-C-22750	Interior surfaces (walls, floors, doors except above and 11 fittings)
		White 27875 or green 24533, conforming to MIL-C-22750	Ceiling
	b. Nonconfigured (Bare shelter)	Lusterless Green 383 (34094), MIL-C-46168 or MIL-C-53039	Exterior surfaces, and doors which may become exterior surfaces during operational use
		Semigloss Green 24533 conforming to MIL-C-22750	Interior surfaces (walls, floors, doors, except above and fittings)
		White 27875 or green 24533, conforming to MIL-C-22750	Ceiling
		NOTES **Markings may be adhesive backed markers or paint conforming to color requirements**	
18.	Unit identification	Lusterless Black 37030, MIL-C-46168 or MIL-C-53039	As prescribed in applicable DA technical publication

NOTE

NOTE 1 Communications-electronic equipment already in use, or that purchased as a nonmilitary item direct from commercial stocks, need not be painted in accordance with this table if the color does not adversely affect equipment operation or the tactical situation. This will hold true particularly in the case of equipment that is housed in shelters, aircraft, vehicles, or vessels.

NOTE 2 Do NOT paint radiating elements, reflectors, radomes, wave guides and insulators until special approval is obtained from the design activity, or unless item(s) are known to have been painted previously with CARC.

Table C-8. Bridging Equipment

ITEM NO	EQUIPMENT	COLOR	PLACEMENT
1.	M48A2 and M60 armored vehicle bridge launchers	Semigloss Green 24533	Interior
		MIL-C-46168 or MIL-C-53039 camouflage pattern	Exterior
2.	Ribbon bridge	MIL-C-46168 or MIL-C-53039 camouflage pattern	
3.	M2 panel bridge (Bailey), Bailey bridge erection equipment, Bailey bridge conversion set, and Bailey bridge cable reinforcing kit	Aircraft Gray 36231, MIL-C-46168 or MIL-C-53039	
4.	Bridge erection set, fixed bridges	Aircraft Gray 36231, MIL-C-46168 or MIL-C-53039, except as follows:	Interior
	a. Parts E159-E-191	Aircraft Yellow 33538	
	b. Parts E36N-E41N, E58N-EG0AN, E122-E129BN and E131N-E133N	Aircraft Red 31136, MIL-C-46168 or MIL-C-53039	
5.	Fixed steel I-beam railway bridge	Aircraft Yellow 33538	
6.	Fixed steel railway bridge, 70 foot	Aircraft Red 31136, MIL-C-46168 or MIL-C-53039	
7.	V-type steel trestle	Insignia Blue 35044, MIL-C-46168 or MIL-C-53039	
8.	Certain parts of the aluminum footbridge structure as outlined in the item specification	Aircraft White 37875, MIL-C-46168 or MIL-C-53039	
		Black conformable non-slip Walkway Compound MIL-W-5044, Type IV, applied while the initial MIL-C-46168 or MIL-C-53039 topcoat is still wet	Walkways

Table C-8. Bridging Equipment-Continued

ITEM NO	EQUIPMENT	COLOR	PLACEMENT
9.	All other bridging equipment	MIL-C-46168 or MIL-C-53039 camouflage pattern in accordance with applicable technical publications **NOTE** **Markings may be adhesive backed markers or paint conforming to color requirements**	Exposed Surfaces
10.	Stenciling and identification	Lusterless Black 37030, MIL-C-46168 or MIL-C-53039	

Table C-9. Other Materiel

ITEM NO	EQUIPMENT	COLOR	PLACEMENT
1.	Towed artillery and multiple rocket launchers	Three-Color MIL-C-46168 or MIL-C-53039 camouflage pattern. Apply identification markings according to MIL-STD-642.	Exterior surfaces
2.	Army materiel intended for field use and not otherwise specified herein	Three-Color MIL-C-46168 or MIL-C-53039 camouflage pattern	Exterior surfaces
3.	Fire control material	Lusterless Green 383, MIL-C-46168 or MIL-C-53039 White 17185, MIL-C-22750, other colors consistent with existing colors Fire control instruments inside vehicles	Exterior surfaces of Fire Control Materiel that are external to vehicles and other field equipment
4.	Conventional and chemical ammunition	Lusterless Green 383, MIL-C-46168 or MIL-C-53039 (34094); other colors according to particular requirements.	Non-tube fired Ammunition and Chemical Ammunition currently painted that do not have special requirements
5.	Chemical warfare	Lusterless Green 383 (34094), except such equipment mounted on vehicles will be painted the same color specified for the vehicle.	
6.	Fire extinguishers (regardless of type, size or location on equipment)	Semigloss Red 21136 in accordance with the military or Federal Specification under which they were procured. Commanders in theaters of operation are authorized to repaint extinguishers in camouflage colors.	Exterior surfaces
7.	Reusable shipping containers/transporters (ammunition packaging, CONEX/MILVAN containers)	Lusterless Green 383 (34094), MIL-C-46168 or MIL-C-53039	Exterior surfaces

Table C-9. Other Materiel--Continued

ITEM NO	EQUIPMENT	COLOR	PLACEMENT
8.	Petroleum distribution equipment and water supply equipment	Three-Color MIL-C-46168 or MIL-C-53039 camouflage pattern	Exterior surfaces
9.	Machine tools and associated shop equipment	Gloss Gray 16187. Exceptions: at fixed facilities such equipment may be painted Semigloss Green 24272. Safety color code markings are in AR 385-30. Emergency stopping switches and bars on such equipment will be painted Gloss Red 11105 or 11136.	Exterior surfaces
10.	Miscellaneous equipment at fixed facilities	See AR 385-30.	
11.	Commercial type items originally procured in nonstandard colors	Maintain in existing colors. When repainting becomes necessary, use appropriate standard colors.	
12.	Equipment used for instructional purposes	Use standard colors except when varied colors may add significantly to the effectiveness of instruction. When equipment is returned to stock, it will be repainted and marked with authorized colors. **NOTE** **Markings may be adhesive backed markers or paint conforming to color requirements.**	
13.	Unit identification		As prescribed in applicable DA technical publications.
14.	Static training equipment		The national symbol and other markings are not required on this equipment

Table C-9. Other Materiel-Continued

ITEM NO	EQUIPMENT	COLOR	PLACEMENT
15.	Aviation Ground Support Equipment	See TM 55-1500-204-25/1	
16.	Reusable Metal Shipping Containers (other than transporters):		
	a. Container that is attached to the weapon system during operation	Three color MIL-C-46168 or MIL-C-53039 system during operation camouflage pattern	Exterior surfaces
	b. All other containers	Lusterless Green 383 (34094), MIL-C-46168 or MIL-C-53039	Exterior surfaces

APPENDIX D
CARC PAINT PLANNING MATRIX

D-1. This appendix contains resource planning information for large painting operations. For certain end items subject to Army CARC and camouflage policy, the matrix furnishes guidelines for estimating required primer/paint quantities and task-hours. The matrix does not cover all Army equipment.

D-1

Abbreviated Item Name and NSN	Primer	Exterior Paint			Interior Paint	Task-hours Required				
		383GN	383BR	Blk		Prep/Final	Int	Ext B/C	CPP	Total
(Example) M-109SPH 2815-01-___-_____	3	3	1	2	4	5	7	2	5	19

1.   Abbreviated Item Name and NSN-Abbreviated name of end item, model number as applicable, and national stock number.
2.   Primer-Total number of gallons of primer required, includes both interior and exterior requirements.
3.   Exterior Paint-Total number of gallons of exterior CARC color required (MIL-C-46168 or MIL-C-53039).
4.   Interior Paint-Total number of gallons of interior paint required (MIL-C-22750).
5.   Task-hours Required-Task-hours required to paint item, stratified as follows:
    a.   Prep/Final-Total task-hours required to prepare item for painting plus total task-hours required after painting (removal of tape, paper, etc.).
    b.   Int-Total task-hours required to paint the interior of the item.
    c.   Ext B/C-Total task-hours required to apply the exterior base coat.
    d.   CPP-Total task-hours required to apply and paint the camouflage paint pattern.
    e.   Total-Total task-hours required (sum of 5a, 5b, 5c and 5d).

Table D-1. CARC PAINT PLANNING MATRIX

Abbreviated Item Name and NSN	Exterior Paint				Interior Paint	Task-hours Required				
	Primer	383GN	383BR	Blk		Prep/Final	Int	Ext B/C	CPP	Total
AH-IG COBRA 1520-00-999-9821	4	4	-	-	1	72	16	19	-	107
UH-1 IROQUOIS HUEY 1520-00-087-7637	3.5	3	-	-	1.5	126	16	22	-	164
OH-6A CAYUSE 1520-00-918-1523	1	1.5	-	-	1	56	4	20	-	80
OH-58 KIOWA 1520-01-169-7137	1	1.5	-	-	1	58	6	16	-	80
UH-60A BLACKHAWK 1520-01-033-0266	4.5	6	-	-	2	224	24	32	-	280
CARRIER APC M113A2 2350-01-068-4077	3	3	1.5	.5	2	4.788	4	7.612	8.39	24.79
CARRIER CP M577A2 2350-01-068-4089	4	4	2	1.5	3	4.788	4	7.612	8.39	24.79
VULCAN TWD M 167A1 1005-01-014-0837	3	3.5	1.5	.5	2	2.788	2	5.612	3.49	13.89
VULCAN SP M163A1 2350-01-017-2113	3	3.5	1.5	.5	2	5.788	4	7.612	8.39	25.79
CARRIER CP 577A1 2350-00-056-6808	4	4	2	1.5	3	4.788	4	7.612	8.39	24.79
INF FIG VEH XM2 2350-01-048-5920	3	3	1	.5	2	9	7	9	18	43
CFV XM3 2350-01-049-2695	3	3	1.5	1	2	9	7	9	18	43

Table D-1. CARC PAINT PLANNING MATRIX
D-3

Abbreviated Item Name and NSN	Primer	Exterior Paint			Interior Paint	Task-hours Required				
		383GN	383BR	Blk		Prep/Final	Int	Ext B/C	CPP	Total
CARRIER M901 1450-00-176-2697	3	3.5	1.5	.5	2	4.788	4	7.612	8.39	24.79
CARRIER CARGO M548 2350-00-078-4545	3	4	2	.5	3	3.289	3	1.777	6.9	14.96
CARRIER M548A1 2350-01-096-9356	3	4	2	.5	3	3.289	3	1.777	6.9	14.96
CHAPARRAL SYS M48A2 1425-01-106-3089	2	3	1	.5	2	5.788	3	4	8.9	21.68
LAUNCHER G/M 1425-01-074-6799	2	2.5	1	.5	-	5.788	3	4	8.9	21.68
CHAP G/M SYS M48 1425-01-069-8877	3	3	1	.5	-	5.788	3	4	8.9	21.68
CARRIER MORTAR M125A1 2350-00-071-0732	3	3	1.5	.5	2	4.788	4	7.612	8.39	24.79
CARRIER MORTAR M106A1 2350-00-076-9002	3	3	1.5	.5	2.5	4.788	4	7.612	8.39	24.79
CARRIER MORTAR M125A2 2350-01-068-4087	3	3	1.5	.5	2	4.788	4	7.612	8.39	24.79
CARRIER MORTAR M106A2 2350-01-069-6931	3.5	3.5	1.5	.5	3	4.788	4	7.612	8.39	24.79

Table D-1. CARC PAINT PLANNING MATRIX

Abbreviated Item Name and NSN	Exterior Paint				Interior Paint	Task-hours Required				
	Primer	383GN	383BR	Blk		Prep/Final	Int	Ext B/C	CPP	Total
COMBAT VEH ITV M901 2350-01-045-1123	3	3	1.5	.5	2	4.788	4	7.612	8.39	24.79
COMBAT VEH ANT 2350-01-103-5641	3	3	1.5	.5	2	4.788	4	7.612	8.39	24.79
FIRST V CONVERSION 2350-01-085-3792	3	3	1.5	.5	2	4.788	4	7.612	8.39	24.79
MLRS	4	5	3.5	1.5	4	9.788	8	11.612	20	49.4
TRK CGO 2 1/2T 2320-00-926-0873	4	7	.5	1	-	13.6	-	12	4	29.6
TRK CGO 5T 2320-00-055-9265	5	8	1	1	-	26	-	12	4	42
TRLR, SEMI 60T 2330-00-089-7265	5	10			-	16.8	-	12	-	28.8
TRLR, SEMI VAN 6T 2330-00-569-9372	4	5			3	26	4	8	-	38
CONTAINER, SHIPPING 8115-01-015-7039	2.5	3			2	6	-	6	-	12
LAUNDRY UNIT 3510-00-782-5294	3.5	7			-	28	-	12	-	40
WTR PURIF (1500) 4610-00-202-6925	4	6			3	58.7	4	20	-	82.7
WTR PURIF (3000) 4610-00-202-8701	6	7			5	61.5	4	20	0	85.5

Table D-1. CARC PAINT PLANNING MATRIX

Change 3 D-5

Abbreviated Item Name and NSN		Exterior Paint			Interior Paint	Task-hours Required				
	Primer	383GN	383BR	Blk		Prep/Final	Int	Ext B/C	CPP	Total
MINE DISP 1095-00-397-3456	5	7	2	1	-	30	-	8	2	40
LUBE UNIT 4930-00-935-4951	2.5	5	-	-	-	28	2	10	-	40
CLOTH REP TRLR MTD 3530-01-017-9124	2.5	5	-	-	-	6.7	-	4	-	10.7
CRANE WHLD 3810-00-043-5354	10	20	-	-	-	80	-	40	-	120
TRLR, SHOP EQ 4940-00-164-2719	5	7	-	-	3	54	4	24	-	82
TRLR, SHOP EQ 4940-01-022-5322	4	5	-	-	3	52	4	24	-	80
M9 ACE 2350-00-808-7100	6	3	1	2	2	107	16	32	16	171
TRK, VAN EXP 2320-00-907-0707	8	12	-	-	4	47	4	24	-	75
TRANSP, RIB BRIDGE 5420-00-071-5321	2.5	5	-	-	-	40.5	-	22	-	62.5
TRLR CGO 1 1/2T 2330-00-141-8050	.5	1	-	-	-	6.7	-	2	-	8.7
FK LFT 4K ELECT 3930-00-327-1603	1	YELLOW 2	-	-	-	12	-	8	-	20

Table D-1. CARC PAINT PLANNING MATRIX

Abbreviated Item Name and NSN	Exterior Paint				Interior Paint	Task-hours Required				
	Primer	383GN	383BR	Blk		Prep/Final	Int	Ext B/C	CPP	Total
CRANE, CRAWLER 12T 3910-00-689-3092	8.5	17	-	-	-	38	-	12	-	50
TRK WRKR 5T 2320-00-055-9258	6	12	-	-	-	30	-	12	-	42
S-280 SHELTER 5411-01-092-0892	2.5	2.5	.25	1.0	2.5	10	2.8	2.8	6	21.6
S-250 SHELTER 5411-00-489-6076	2	2	.5	1	2	10	2.3	2.5	4.5	19.3
1 1/2 TRAILER M998	1	1	.5	1	N/A	64.30	N/A	2.5	3.5	70.3
400 GL WATERTANKER M149A2	1.5	1.5	.5	1	N/A	68	N/A	2.5	3.5	74
5T TRUCK M923 2320-01-050-2084	3	3	1	2.5	N/A	20	N/A	3.5	8.5	28.85
1 1/4T TRUCK UTIL M-105 CUCV	1.5	1.5	1	2	N/A	81	N/A	2	6	89
MRP003A GENERATORS 6115-00-465-1030	1.8	1.8	1	2	N/A	13	N/A	3	5.5	21.5
GEN 200 KW 60 HZ 6115-00-133-1904	1.5	1.5	.5	1	N/A	32	N/A	2	3.5	37.5

Table D-1. CARC PAINT PLANNING MATRIX

Abbreviated Item Name and NSN	Exterior Paint				Interior Paint	Task-hours Required				
	Primer	383GN	383BR	Blk		Prep/Final	Int	Ext B/C	CPP	Total
PU-650 6115-00-258-1622	3	3	1	2.5	N/A	2.9	N/A	4	7	13.9
MJQ-10 6115-00-394-9582	3	3	1.5	2.5	N/A	5.87	N/A	4	7	16.37
V-528/T CABLE REEL TRAILER 2330-01-141-6330	1.5	1.5	.5	1	N/A	5.81	N/A	2.5	3.5	11.81
AN/MJQ-15 6115-00-400-7591	3	3	1	2	N/A	5.87	N/A	4	7	16.87
ANTENNA TPN-18 5840-00-944-2452	.5	.5	.25	.5	N/A	12	N/A	1.75	2.5	16.25
AN/VS5-3 SEARCH LIGHT 5855-00-405-0404	.25	.25	.12	.12	N/A	2	N/A	.75	1.5	4.25
RAPID COMPUTER VAN 5820-SF-P83-33T	6	6	2	5	5	50	8.0	12	16	86
2 1/2T TRUCK M35A2	2.5	2.5	.25	1.5	N/A	36	N/A	3	6.5	45.5
PU-753/M 10KW 60HZ 6115-PU753	2	2	.5	1.5	N/A	5	N/A	3.5	5.5	14
MUST WARD CONTAINER 5410-00-809-6634	3	3	1	2.5	3	20	3	3.5	7	33.5
AN/MSC-25 5895-00-021-2088	5	5	1.5	3.5	5	213	4	10	14	241

Table D-1. CARC PAINT PLANNING MATRIX

Abbreviated Item Name and NSN	Exterior Paint				Interior Paint	Task-hours Required				
	Primer	383GN	383BR	Blk		Prep/Final	Int	Ext B/C	CPP	Total
PU-732/M 15/400 6115-PU-732/M	3	3	1	2.5	N/A	2.93	N/A	4	7	13.93
GEN ST TM 5 KW PU620 6115-PU620GEN	2	2	.5	1.5	N/A	4.12	N/A	3.5	5.5	13.12
PU-405/A MGEN 15/600T 6115PU405GEN	3	3	1	2.5	N/A	2.93	N/A	4	7	13.93
GENST 3 KW PU-625 6115PU625	2	2	.5	1.5	N/A	6.51	N/A	3.5	5.5	15.51
GST DIES MJQ-18 6115ANMJQ18	1.5	1.5	.5	1	N/A	5.25	N/A	2.5	4.5	12.25
GEN ST KW PU751 6115PU751	2	2	.5	1.5	N/A	3	N/A	3.5	5.5	12
M60A3 2350-01-061-2306	5	17	1.5	1.5	5.5	32.63	13.95	12.78	13.5	72.86
M728 2350-00-795-1797	5.5	18.5	2	2	5.5	41.57	18.98	17.82	14.5	92.87
M48A5 2350-01-059-1504	5	17	1.5	1.5	5.5	32.63	13.95	12.78	13.5	72.86
M88 2350-00-122-6826	5.5	18.5	2	2	6	42.01	19.2	17.15	14.5	92.86
M1 ABRAMS 2350-01-061-2445	6	18	1.5	1.5	6	72	14	15	13.5	114.5
M551A1 2350-00-140-5151	4.5	13.5	1	1	4	52	10	10	11.5	83.5

Table D-1. CARC PAINT PLANNING MATRIX

D-9/(D-10 blank)

GLOSSARY

The following terms are defined as they are used with respect to painting and related operations.

Abrasive resistance - This property is comparable to toughness rather than hardness. It is that property exhibited by the surface of a paint, enamel, or varnish which will resist being worn away by rubbing or friction.

Adhesion - As used in reference to paint films, adhesion is the tendency of the film, when dry, to bond to the surface upon which it has been applied.

Alligatoring - Rupturing of the top paint coat, which causes the surface to break up into irregular areas separated by wide cracks in an "alligator hide" fashion.

Atomization - A paint and air mixture, whose round or oval pattern is generated by the mixing of paint/material and compressed air at the air cap of a spray gun.

Binder - The nonvolatile portion of a paint vehicle. Binders may be drying oils, resins, or a number of other substances such as casein, chlorinated rubber, nitrocellulose, or ethyl cellulose.

Blast cleaning - Blast cleaning to "white metal " is defined as blast cleaning which removes completely all visible mill scale, rust, paint, foreign matter, and pitted areas from the surface of the metal. The end result must be a light-gray steel surface of uniform appearance.

Bleeding - When the color of a pigment in a previous coat comes through the topcoat. This usually occurs when a previously applied pigment is soluble in the medium of the newly applied topcoat. Asphalt and colored resins may also bleed.

Blistering - A condition in which the paint coat is detached and raised from the surface upon which it is applied as the result of gases or liquids, usually water, forming beneath the coating.

Blushing - The precipitation of ingredients of a paint film when it dries, which may be caused by condensation of moisture on the film or by improper composition of the paint.

Body - A paint is said to have "body" or to be "bodied" when it is thickened above its normal condition. Thus the "body " of a paint is its relative thickness. The degree of "body" is in proportion to a paint's viscosity.

Boxing - The process of mixing paint by pouring it back and forth from one container to another.

Brightness - The brightness of a paint film is measured by the percentage of incident light reflected from the film.

Brushing property - The quality a paint displays when it is applied to a surface, as affected by its viscosity, mobility, consistency, composition, etc.

CARC - Chemical Agent Resistant Coatings; a system of primers and topcoats that are required on all combat, combat support, and combat service support equipment. CARC is used to provide camouflage protection and/or chemical agent resistance to liquid chemical agents. Since CARC does not absorb chemical agents it does not create long term contact hazards.

Catalyst - A substance used in the manufacture of paint that causes a chemical and/or physical reaction to take place.

Chalking - When loose powder, which can be removed by gentle rubbing, appears on the paint film or just beneath the surface. A good quality paint applied correctly should chalk very slowly. Chalking should be a

Glossary 1

gradual process over a period of years, so that when repainting becomes necessary, the surface is in good condition to receive the new coat, with little, if any, preparatory work required.

Checking - A paint film condition with slight breaks in the film surface, causing the undercoats to be visible.

Coat-Coating - A protective film of paint, varnish, primer, lacquer, etc.

Confined Space - Any area where dilution ventilation cannot take place, or where air flow is obstructed. Refer to para 1-7b for examples of confined spaces.

Cracking - Breaks in a paint film which extend through the film to the underlying material.

Crawling - Creeping-The collection of paint into little drops or islands on the applied surface.

Drying oil - An oil which, when exposed in a thin film to the air, possesses to a marked degree the property of readily absorbing oxygen and changes to a relatively hard, tough, and elastic substance.

Dulling - The loss of gloss which develops in a varnish film after drying.

Enamel - A paint which has the ability to form an especially smooth film. An enamel always contains pigment and has moderate hiding power and color. Some enamels dry to a flat or eggshell finish instead of a gloss finish. An enamel is a finish that comprises a dispersion of pigments in a varnish or resin vehicle or is a combination of both. This includes all CARC coatings. Enamels dry by a process of oxidation and/or polymerization.

Feathering - The procedure of thinning a coating between a bare and a painted surface by sanding to a fine edge. It is used when preparing touchup spots for painting and where an invisible lap is required.

Finish system - A particular combination of primers, topcoats, and pretreatment materials that are used on a specific type of surface in order to obtain a desired result (i.e. camouflage, chemical agent resistance, etc.) Also referred to as a paint system.

Filler - A special paint used for filling pores or other breaks in a surface to make it smooth for further painting. When applied and exposed to the air, a filler should dry to a relatively hard, permanent solid, capable of supporting subsequent coats.

Flaking - When small pieces of the paint coat fall away.

Gloss - The degree of mirror-like reflection of a painted surface.

Hiding power - The ability of a paint or paint material to cover up a surface so that the surface cannot be seen.

Hydrocarbons - An organic compound, such as acetylene or benzene, that contains only carbon and hydrogen, and occurs in petroleum, natural gas, coal, and bitumens.

Induction - A period of time required for recently mixed materials to begin to react prior to use.

Lacquer - A clear or pigmented finish whose vehicle is cellulosic or phenolic, with or without other resins or plasticizers. Lacquer dries by solvent evaporation.

Leveling - The ability of a paint to flow, leaving a smooth film when brushed onto a surface.

Mildew - A fungus frequently noted on surfaces exposed to damp, warm climates. This is usually found on surfaces covered with paint of a soft nature. Such paints act like flypaper and afford lodging for windblown matter from decayed and dried vegetation. Sometimes the oil with which the paint is made or mixed from is infected and offers a breeding place for mildew spores.

Opacity - The degree of obstruction of a coating to the transmission of visible light.

Oxidation - In coatings, the curing reaction which requires oxygen from the air to form the film.

Glossary 2

Paint - Paint is composed of a pigment and a vehicle. The pigment, or solid component, is dispersed in a paint, provides color to the paint, and enables it to form a film on the painted surface. T he vehicle is the liquid portion of a paint, which includes components that serve as binders, as well as volatile components known as thinners. The binder portion of the vehicle, like the pigment, is film forming. After evaporation of the volatile content, drying is by oxidation.

Paint system - The protective paint barrier that covers a painted object, and may consist of a pretreatment coat, primer coats, intermediate coats, and/or finish or top coats. Also referred to as a finish system.

Peeling - A more aggravated form of scaling, usually due to the presence of moisture when the paint was applied or to faulty application of the priming coat.

Pigment - The fine, solid particles used in the preparation of paint, substantially insoluble in the vehicle. Pigments provide coloration, corrosion resistance, strength, hardness, increased durability, and control of gloss.

Polymerization - The reaction, usually at elevated temperatures, in which two or more components of the substance combine to form a more complex molecular structure, which has the property of curing or solidifying with or without the absorption of oxygen.

Pretreatment coat - The wash primer or preprimer paint film that is applied under the regular primer paint coat and is used for better bonding and corrosion control.

Primer - A paint which is intended for use as the initial covering for a surface and is usually followed by other coats, often of a different type of paint. Primers are also called undercoats, and usually contain corrosion resistant properties.

Respiratory protection, approved - Approved respiratory protection equipment is that equipment tested and listed as satisfactory according to standards established by a competent authority, such as the National Institute for Occupational Safety and Health (NIOSH), or the Mine Safety and Health Administration (MSHA), to provide respiratory protection against the hazard for which it is designed. The specific approval authority may be specified by law for particular hazards.

Runs - Sags - Irregularities of the paint film due to uneven flow of the paint.

Scaling - Flaking of the paint film in an aggravated form in which the paint coating falls off in large sections.

Solvent - A volatile thinner, particularly for varnishes and lacquers.

Spotting - The appearance of discolored spots on a painted or varnished surface.

Spray coat - A spray coat consists of one or more coats, depending on the paint, and should be considered as that amount of paint applied at one time, just short of sagging, running, or wrinkling.

Stripper - Any solution used for paint removal.

Stripping - The process of removing paint from a painted surface by means of a stripper.

Sweating - A term used to describe the reappearance of luster on a varnished surface which has been rubbed to a dull finish.

Thinner - Thinners make a paint workable, adjusting the consistency for easy application, and producing a uniform film that will penetrate and adhere to the surface. The thinner, being volatile, evaporates and does not provide part of the dried surface film.

Toxic - A paint or other product that has poisonous qualities. While some paints and related materials have toxic qualities with respect to the using personnel, products which are named "toxic paints" are developed for their poisonous qualities against fungi, teredo, barnacles, etc.

Varnish - An unpigmented (clear) finish whose vehicles consists of resins and both drying and non-drying oils. After evaporation of the volatile content, drying is by oxidation and/or polymerization.

Vehicle - The liquid portion of a paint which carries the pigments. Anything that is dissolved in the liquid portion of a paint becomes a part of the vehicle.

Washing - Paint films sometimes allow the pigment to "wash" out under action of the elements. When rubbed, a wet, soapy, emulsion will be formed. This is termed "washing".

Wrinkling - Sometimes referred to as "crinkling", "puckering", or "crimping", this describes a condition in which the paint film gathers in wrinkles. It frequently occurs when paint or varnish is applied at low temperatures.

Glossary 4

INDEX

Index 4

By Order of the Secretary of the Army:

JOHN A. WICKHAM, JR.
General, United States Army
Chief of Staff

Official:

R. L. DILWORTH
Brigadier General, United States Army
The Adjutant General

DISTRIBUTION:
To be distributed in accordance with DA Form 12-34B, Maintenance requirements for Painting Instructions for Field Use

The Metric System and Equivalents

Linear Measure

1 centimeter = 10 millimeters = .39 inch
1 decimeter = 10 centimeters = 3.94 inches
1 meter = 10 decimeters = 39.37 inches
1 dekameter = 10 meters = 32.8 feet
1 hectometer = 10 dekameters = 328.08 feet
1 kilometer = 10 hectometers = 3,280.8 feet

Weights

1 centigram = 10 milligrams = .15 grain
1 decigram = 10 centigrams = 1.54 grains
1 gram = 10 decigram = .035 ounce
1 dekagram = 10 grams = .35 ounce
1 hectogram = 10 dekagrams = 3.52 ounces
1 kilogram = 10 hectograms = 2.2 pounds
1 quintal = 100 kilograms = 220.46 pounds
1 metric ton = 10 quintals = 1.1 short tons

Liquid Measure

1 centiliter = 10 milliters = .34 fl. ounce
1 deciliter = 10 centiliters = 3.38 fl. ounces
1 liter = 10 deciliters = 33.81 fl. ounces
1 dekaliter = 10 liters = 2.64 gallons
1 hectoliter = 10 dekaliters = 26.42 gallons
1 kiloliter = 10 hectoliters = 264.18 gallons

Square Measure

1 sq. centimeter = 100 sq. millimeters = .155 sq. inch
1 sq. decimeter = 100 sq. centimeters = 15.5 sq. inches
1 sq. meter (centare) = 100 sq. decimeters = 10.76 sq. feet
1 sq. dekameter (are) = 100 sq. meters = 1,076.4 sq. feet
1 sq. hectometer (hectare) = 100 sq. dekameters = 2.47 acres
1 sq. kilometer = 100 sq. hectometers = .386 sq. mile

Cubic Measure

1 cu. centimeter = 1000 cu. millimeters = .06 cu. inch
1 cu. decimeter = 1000 cu. centimeters = 61.02 cu. inches
1 cu. meter = 1000 cu. decimeters = 35.31 cu. feet

Approximate Conversion Factors

To change	To	Multiply by	To change	To	Multiply by
inches	centimeters	2.540	ounce-inches	newton-meters	.007062
feet	meters	.305	centimeters	inches	.394
yards	meters	.914	meters	feet	3.280
miles	kilometers	1.609	meters	yards	1.094
square inches	square centimeters	6.461	kilometers	miles	.621
square feet	square meters	.093	square centimeters	square inches	.155
square yards	square meters	.836	square meters	square feet	10.764
square miles	square kilometers	2.590	square meters	square yards	1.196
acres	square hectometers	.405	square kilometers	square miles	.386
cubic feet	cubic meters	.028	square hectometers	acres	2.471
cubic yards	cubic meters	.765	cubic meters	cubic feet	35.315
fluid ounces	milliliters	29,573	cubic meters	cubic yards	1.308
pints	liters	.473	milliliters	fluid ounces	.034
quarts	liters	.946	liters	pints	2.113
gallons	liters	3.785	liters	quarts	1.057
ounces	grams	28.349	liters	gallons	.264
pounds	kilograms	.454	grams	ounces	.035
short tons	metric tons	.907	kilograms	pounds	2.205
pound-feet	newton-meters	1.356	metric tons	short tons	1.102
pound-inches	newton-meters	.11296			

Temperature (Exact)

°F   Fahrenheit   5/9 (after   Celsius   °C
     temperature subtracting 32) temperature

PIN: 014917-000

www.ingramcontent.com/pod-product-compliance
Lightning Source LLC
Chambersburg PA
CBHW080049280326
41934CB00014B/3256